IAN WALTERS

I0486602

Statistical
Anthropology

outskirts
press

Statistical Anthropology
All Rights Reserved.
Copyright © 2018 Ian Walters
v2.0

The opinions expressed in this manuscript are solely the opinions of the author and do not represent the opinions or thoughts of the publisher. The author has represented and warranted full ownership and/or legal right to publish all the materials in this book.

This book may not be reproduced, transmitted, or stored in whole or in part by any means, including graphic, electronic, or mechanical without the express written consent of the publisher except in the case of brief quotations embodied in critical articles and reviews.

Outskirts Press, Inc.
http://www.outskirtspress.com

ISBN: 978-1-4327-9147-6

Cover Photo © 2018 thinkstockphotos.com.. All rights reserved - used with permission.

Outskirts Press and the "OP" logo are trademarks belonging to Outskirts Press, Inc.
PRINTED IN THE UNITED STATES OF AMERICA

This book is dedicated to the memory and legacy of Franz Boas, founder and father of statistical anthropology.

Contents

Preface

In a strange yet meaningful way science seems to divide modern anthropology into enemy camps. On the one hand sits the moiety of practitioners of scholarship that deals in hypotheses, numbers and the hardness of objects: biological anthropology and a large part of archaeology. On the other sit the practitioners from the moiety of the post-modern, cultural studies-influenced ethnography, where imaginaries are unpacked, tropes are traded and spaces negotiated.

Museum objects and their indigenous origins perhaps help locate some contemporary material culture studies in this latter camp too. But this is given the caveat that those studies for all the early years of the profession of anthropology located themselves in the science camp. Where anthropological linguistics is located I have no idea. In the past some linguists, after the manner of the ethnologist Levi-Strauss – who made productive use of language and linguistic principles - saw themselves as being related to if not sprung forth from science and mathematics. Perhaps now they, the linguists, merely desire to be elsewhere. Old fashioned ethnology is already elsewhere; like Dunbar it has been disappeared.

The book does not seek to malign the majority of anthropology which gets by perfectly adequately with analysis by narrative and reporting, interpreting and concluding via observation. Words usually communicate our evidence. Most of our classic literature is of this form. I merely aim in what follows to lay down some principles, via short examples, for studies where numbers are involved and statistical inferences are to be drawn from numerical data.

So from the above we could see science as either something you take for granted, as the first moiety would see it, and practised as part of daily fare, or it is some abomination that humanities can neither embrace nor abide, as many post-modern ethnographers would see it. This book presumes neither, but does in an admittedly strong sense hanker back to the days of scholars like Boas, Radcliffe-Brown and Levi-Strauss who were convinced that what they were about was no less than the science of man. Consequently it blatantly attempts to make the case for seeing value - in regard to many aspects of anthropology - for a continuation of a scientific approach.

I give short introductions to examples of some of that work, ancient and pedigreed, juxtaposed with more modern examples of how such mathematical and scientific thinking can still constitute a revealing, insightful and important component of our anthropology.

━━◦《◉》◦━━

The Frontispiece image of Franz Boas was downloaded from among the various online pictures of him and relating to him. These are available freely and readily on the internet.

The Australian Bureau of Statistics (ABS) graph is downloaded under a Creative Commons Attribution 2.5 Australia whereby I am able to "copy and redistribute" the material included on the ABS website.

I am grateful for being able to use the map originally drawn by Edwin Ferdon Jr which appeared in Elman Service's *Cultural Evolutionism*

(1971:129), published in 1971 by Holt, Rinehart & Winston. All attempts to contact the publisher have led to nought, with my contact email returned from a dead letter box while no website or home page seems to now exist.

Professor Christopher Bronk Ramsey kindly provided me with an electronic offprint of his paper from *Radiocarbon*.

Professor Sean Ulm kindly provided me with an electronic offprint of the Veth *et al.* Barrow Island paper in *Quaternary Science Reviews*. In addition Mr Kane Ditchfield kindly provided me with details of his OxCal Bayesian analysis from the Appendix proofs of that same paper.

At Outskirts Press I thank Pat Wilson for manuscript review and Lisa Jones for overseeing the MS through production.

One

Introduction

As the title makes evident, this little book aims to address analyses of various topics in anthropology using a statistical approach. My objective is to demonstrate for readers, via discussion of a range of simple examples which illustrate my theme, that statistical thinking – which usually means thinking about the behaviour of large number cohorts (Tolman 1979:71) – can provide useful and important insights into interpretation of patterns of the cultural past or the social present. To the best of my knowledge no anthropological writer has attempted this before. Never mind, we are only just over a century off the pace. For it was as recent as the very early years of the twentieth century that Josiah Willard Gibbs, the American mathematical physicist, and Albert Einstein, the German Jew working in Switzerland, independently invented what came to be known as statistical physics.

Gibbs, that "great and modest scientist" (Tolman 1979:14), the "man of genius" sometimes referred to as the "Maxwell of America" (Nahin 2002:188), emerges as a singular giant of statistical thinking in the physical sciences. He appears to have worked mainly alone, being rather removed from the mainstream of theoretical physics which was, at that time, very much gravity-centred on Germany. In fact in Einstein's first publication on the topic he failed to cite Gibbs' precedence, as he had not yet heard of the American. Courtesy of colleagues who pointed him in the right direction, he later acknowledged Gibbs in further work, saying that had he known at the time he would not have published in the area - except for a few comments (Pais 2008:55). In fact he also said of Gibbs, who he never met, that he "might have placed him beside Lorentz" (Nahin 2002:187) – coming from Einstein that would have been the highest tribute.

As an aside I find this ironically humour-laden, for Einstein, one of my serious culture heroes (Walters 1980, 2014), was known to be somewhat sensitive about his precedence on many issues and ideas; scribbling off petulant missives to researchers who had published or pontificated on some given topic without acknowledging the great man's priority (Galison 2003:77-8; Penrose 2008:ix). Yet when it came to being diligent about the precedence of others he was, shall we say politely, rather less energetic and blithely less concerned.

The precedence issue was ill-founded anyway, for the real beginnings of statistical thinking in mathematical physics began over half a century before with another who Germany claimed, the Austrian Ludwig Boltzmann, also in the pantheon of my esteemed culture heroes. Einstein was the one who really put terminal nails in the coffin of thought that denied molecular (or atomistic) thinking in physics, but it was Boltzmann who carried the torch for the atomistic view of the material world for many decades previous. His reinterpretation of the Second Law of Thermodynamics using a statistical approach gave science many fruitful insights into that law, and to dealing with its nonpareil consequence, entropy. I will say more about all this in the following Section of the text, where, as you will see, yet another great culture hero Franz Boas, takes pride of place.

What the book aims to do is this: to deploy the methods of statistical thermodynamics (Schrödinger 1989), statistical mechanics (Tolman (1979), statistical physics (Einstein 1998a, 1998b), or statistical ecology (May 1975) – take your pick – to show how these methods, and this way of thinking, can be useful for approaching a certain set of anthropological problems. Not all anthropological problems by any stretch are amenable to such an undertaking. In fact in one major branch of our discipline, ethnography, there exists a traditional and living denial of the need for or applicability of large number cohorts. Ethnographers focus on individuals, and perhaps their family or friends. But ethnologists on the other hand, seek generalisations using comparative ethnographic materials. Here some branches of the trade are and have long been considered amenable to statistical analysis, as I will show. As are all other branches of our diverse discipline, where problems arise that can be effectively dealt with using large samples, of people, of names, of bones, or of artefacts. These in turn mostly involve data that either fall naturally, or are readily organised into, classes or categories. But first it's probably beneficial to say something for those unacquainted with our diverse discipline.

A short digression into the structure of anthropology

Before embarking on my examples given in sections 3 through 9, I include a digressionary section on the academic structure of my profession. The aim of this is to show those who may not be familiar with it, or are learning it at undergraduate level that anthropology is a very diverse and broad-ranging discipline. So, as my examples outline, we may be thinking, statistically or otherwise, about issues and questions that begin with the earliest hominids from 6 million years ago, to more recent skeletal problems to do with human evolution and cultural development, to issues involving archaeological settlement patterns among some of our more recent ancestors, to the makeup of economic resource allocation in various societies, past or present, or even to the patterns present in social organisation – along with its linguistic and kinship terminology - of colonially encountered indigenous peoples, and ending with some thoughts on the social and economic diversity encountered in modern global society. Self indulgently I even sneak in a short example from my own doctoral thesis – on the makeup of hunting catches and the interpretation of these - from all those decades ago (Walters 1987), to show another aspect of this kind of thinking and the insights that can be induced.

Anthropology is the study of people, or more grandly, the study of humankind. Anthropology is unique among the areas and disciplines of human scholarship in that it alone studies the full spectrum of people's lives from the very physical through to the most symbolic and abstract, as well as from the deepest past of our history through to the cutting edge present, including making tentative suggestions for policies and strategies concerned with the future – possibly catastrophic - directions of our lives.

It covers human prehistory, human evolution, archaeology, physical anthropology, material culture, anthropological linguistics, ethnography and ethnology - or as these last two are usually known: cultural and social anthropology. Anthropologists research people and their cultures at all periods in their history, from the deep past through to contemporary times, from all parts of the world, in all types of societies.

Anthropologists work with people of different cultures and societies in an attempt to understand others, to understand ourselves, and hopefully make the world a better place for all of us. Anthropology examines peoples in their cultures in the past, in the present, so we can have things to say about how it might be in the future.

Social and cultural anthropologists study the ways people live in contemporary situations, how they relate to each other, what their families are like, how they make a living, how they govern themselves, what laws they make and why, what belief systems they have, who has power over their lives. Such study is called *ethnography*, or *ethnology* - as it used to be and is still sometimes known - and professionals who do this kind of study are known as ethnographers.

Anthropological linguists study human languages in all their forms, past and present. They study how people communicate with each other in social situations, what can be said and what can't, why this is so, what influences language diversity and change.

Archaeologists study the human past. They examine the artefacts or material culture left by peoples from times past in order to understand lifeways, settlement patterns, how people made a living, their migrations, trade and gift exchange.

Physical anthropologists or *biological anthropologists* study the evolution and development of humans from earliest times to the present, examining variation in body form, why and how people are different, what caused biological varieties to give rise to the peoples of today. They mostly study bones from fossil or archaeological sites, but they can also study soft tissue characteristics too, such as blood groups or even mitochondrial DNA patterns.

In general anthropologists study people and their cultures, past and present. For it is anthropology that has demonstrated, for over a hundred years now, that culture is the central lynchpin for understanding human behaviours, actions, objects and beliefs. Edward Burnett Tylor (1832-1917), the founding father of British anthropology, gave us the most famous definition of culture:

> Culture, or civilization, taken in its wide ethno-graphic sense, is that complex whole which includes knowledge, belief, art, morals, law, custom, and any other capabilities and habits acquired by man as a member of society.
>
> Tylor (1871:1)

It is the complex whole that envelops us, it is the webs of signifi-cance within which we exist and give meaning to our lives. It is what we know, what we believe, and the created institutions of our lives such as our laws and our morals, it is our other creative achievements such as the arts and customs. It is what we learn as children that makes us behave the way we do, knowing what is right and what is wrong, who is who, and which way is up.

For most of the twentieth century culture was seen as learned, shared, integrated, and in an evolutionary sense, adaptive. Clifford Geertz (1973) taught us that culture was semiotic, embroiling us in its webs of significance that are of our own making. We can at best, try to interpret culture in order to gain comprehension of what oth-ers are up to.

The special case of ethnography & ethnology

Post-modern writers like James Clifford (1988) take us on from this to contemporary ideas of culture and its study by anthropolo-gists, and in particular, ethnographers. A culture is an "open-ended, creative dialogue of subcultures, of insiders and outsiders, of diverse factions" (Clifford 1988:46). Culture is not a tradition to be saved but is "assembled codes and artifacts always susceptible to critical and creative recombination" (Clifford 1988:12). Contemporary interpretive anthropology views cultures as "assemblages of texts" (Clifford 1988:41). Some of us would see it to be simply the interpretation of people and what they do, think and say. Whichever, we need to consider the special case of ethnography with ethnology, as it seems to disrupt my agenda for a scientific statistical anthropology.

Ethnography then, is the interpretation of cultures (Clifford 1988:39). It is necessary to see ethnography as a "constructive negotiation involving at least two, and usually more, conscious, politically significant subjects" (Clifford 1988:41). Ethnography is "an explicit form of cultural critique" (Clifford 1988:12). Contemporary academic ethnography is not "interpreting distinct, whole ways of life," but is a "series of specific dialogues, impositions, and inventions" (Clifford 1988:14). Cultural difference "is no longer a stable, exotic otherness; self-other relations are matters of power and rhetoric rather than of essence" (Clifford 1988:14).

Doing ethnography is intervention, in a world not of your own. It is an attempt to understand, to comprehend a Martian universe. It is interpretation, and it has its own problematic dimensions. "Intervening in an interconnected world, one is always, to varying degrees, 'inauthentic': caught between cultures, implicated in others" (Clifford 1988:11). A sense of difference or distinctness "can never be located solely in the continuity of a culture or tradition. Identity is conjunctural, not essential" (Clifford 1988:11). Clifford shares with Edward Said a "search for nonessentialist forms of cultural politics" (Clifford 1988:11).

That's all very well for modern or post-modern political correctness. But like all of anthropology, ethnography's roots lie in somewhat darker motivations. For this was the discipline of colonialism and global European conquest. It was helpful for colonialists to know about the natives whose lands and treasures they were usurping and purloining. Native social structures, ecological rounds, belief systems were all of tactical and strategic assistance to European powers intent on forcing their wills upon native lands, wealth and property. Images of Royal Flying Corps Sopwith Camels bombing and strafing the cattle herds of Sudanese Nuer and Dinka springs to mind (Kuper 1983:88,110). These also go some way to explaining the Nuer coldness to their new visitor Evans Pritchard a few years later. Or speaking of wealth and property, of John Batman acquiring the land of what is now Melbourne for his own future Gotham City from the traditional owners, by handing over some beads and associated trinkets.

In addition as something of an aside we must not be neglectful of the holy true believers whose proselytising representatives arrived in train with the military and political powers, to bring the heathens out of their own purgatory by introducing them to ours. Ethnography had a role here too, as did linguistics. Not partnership, but certainly not as opponents either. Everywhere the Europeans went their Bible was translated into some form of local language or lingua franca. Their god would prevail come hell or high water. Many ethnographers were not implicated in this thank goodness, and often we see from comments that relations were not always that rosy with the missionaries.

Ethnology was what most of these early practitioners claimed to practise. Ethnology is plural and global. Ethnography as is clearly demonstrated above by post-modern thinkers, is singular. In delicious part my argument in this book is for a strong focus to return to ethnology: to its intimate and valuable relations with all other branches of anthropology including modern applied aspects such as heritage studies and land title research. It can be nonessential, it can be conjunctural, it can deal with dialogues and inventions. But it can still pursue generalisations, and even law-like ones at that. It can pursue rigorous theory that is predictive but which contains within itself consequences of practicality. It can be scientific.

Back to statistics and this book

Anthropological readers will be relieved to know one thing: despite its title, this is a mathematical book that contains no mathematics. Well, not quite. It is in fact a mathematical book chock full of mathematics. However, to skim or flip through the pages you wouldn't twig to that. Which is to say there is no sign of mathematical notation, there are no equations and no formulae. It is not a cook book or primer that tries to teach you how to calculate means and standard deviations. For none of that is relevant to my argument really. If the desire takes you to explore the nuts and bolts of the mathematics, you can look up other works, some of which I reference, to investigate the equations, their derivations, the symbolism and notation. Or you can do what generations of anthropologists and other social scientists have done: open your copy of SPSS and click on your choice of buttons, whereupon your calculations are done for you whether you understand them or not.

Here in this work I simply want to get across the ideas of statistical thinking, their usefulness, their pedigree, unencumbered and uncluttered. I do this via words, plain and almost always jargon-free. And I do it by using examples, most of them quite short.

A minor issue for me was how to structure the sections. Chronological order of published appearance seemed logical, but did not fit well with the development of the theme I am trying to draw out. So following the introduction of some fundamental laws and the anthropology of Franz Boas, I have opted for the next few Sections (4 – 6) to follow from the foundations of a statistics-based order of presentation, using examples to do with frequency distributions, normal curves and the like. This culminates in the idea of transformations in such distributions, and how they can be used to characterise social and cultural change. Especially so is this for dramatic or catastrophic change such as witnessed in the very big hierarchy societies of the Holocene.

Then follow two Sections (7 – 8) that deal with categories, classes or species. Much of large number data falls naturally or taxonomically into such groupings. Hence the need for their consideration. Ways to deal with the distributions that follow from them to produce insightful outcomes is and has been of interest in many social and cultural contexts. This holds whether it is research dealing with classic studies like Radcliffe-Brown's analysis of Australian social organisation (Radcliffe-Brown 1931), or my own recent work on a species concept for insights into the industry of sex workers (Walters 2013).

The following Section 9 can be thought of as part of a build up to a climax, dealing as it does with the massive inequality of the modern world and the way capitalism is revealing itself as about little more than the acquisition and hoarding of wealth by the so-called 1% at the socio-economic top of global industrialised society while driving down and exploiting everybody else. Statistical anthropology here can be broadened by the contributions of and relevance to economists, politickers and sociologists.

Section 10 shows how statistical thinking can be combined usefully with probability and decision making under uncertainty to produce strong outcomes of likelihood even when samples are smallish. Based on the central idea of Bayes' Theorem (named after its discoverer the nineteenth century mathematician Thomas Bayes) I give two recent examples (one from archaeological dating, the other from ethnology) that demonstrate the huge value than can be extracted from what is known as Bayesian inference, or Bayesian analysis.

Finally in Section 11 I offer my *piece de resistance*: I move against the tide of history and perhaps popular academic opinion – at least in some quarters of anthropology - to hark back to days gone by when science was a goal of our profession. I propose a new law based on two classic studies from the early and mid-twentieth century respectively. One of these studies hails from ethnology, the other from population genetics.

Why would I do this? Just to show we can, that's why.

Notes to Section 1:

Hendrik Lorentz, Nobel Laureate 1902, was possibly the only scholar – fellow student friends apart – who exerted a significant personal influence on the young Einstein. Certainly he was the only person who qualifies for what might be seen as approaching the role of a mentor for the great man. They corresponded, Einstein went to Leiden and met with the Dutchman, Einstein relied on the Lorentz transformations to play a crucial role in his theory, and they met again at the first Solvay Conference of 1911 when the tyro was invited as one of the young guns while the older man served as the Chairman. Lorentz transformations played a crucial part in the development of the special theory of relativity, so much so that for the first decade or so after the 1905 publication of Einstein's great paper, the theory was widely known in physics circles as the Lorentz-Einstein theory. For AE to liken Gibbs to Lorentz was indeed indicative of the very highest regard in which he held both men.

Two

These Fundamental Laws

"The fundamental laws which governed the growth of culture and civilization seem to manifest themselves conspicuously, and the chaos of beliefs and customs appears to fall into beautiful order."

(Boas 1974b:107)

Franz Boas was the most intellectually radical anthropologist who ever lived. Trained in his native Germany in physics (born 1858), he branched into that new set of developments known as psycho-physics, a pursuit he shared with such luminaries of German (or Austrian) science as Nobel laureate Erwin Schrödinger (Moore 1989:43), which led him eventually into a doctoral thesis on the colour of sea water. This done he embarked in 1883 upon a fifteen month sojourn of fieldwork in Baffin Land among the Inuit. He had moved in later undergraduate work into geography, so perhaps he would have described his work at that time as some kind of geographical psycho-physics. However it quickly evolved into the cross cultural study of people, their manner of making a living, and their ways of seeing the world. This work can be seen as inspired, or at one with, some findings on observers and observations, which he had originally incorporated into his PhD dissertation. All this would see him settled eventually into the developing science of anthropology.

Thus did Boas' career in the academy take off jerkily, slowly, as he then migrated to the USA where he serially took positions at the journal *Science*, the Museum of Natural History, Clark University, before gaining a job at Columbia University in New York City in 1896. He became Professor of Anthropology (within the Psychology Department) in 1899. Then when a Department of Anthropology was established in 1901, he was invited to head it up (Boas 1974d:290; Stocking 1974c:284). There he would remain until his death, in retirement, on 21 December 1942, building an academic empire as the leading light, head, mentor, patron, and inspiration in what became the world's leading anthropology department. Only Boas, the icon, the legend, could conspire to leave us on the above date during a Faculty Club luncheon, dying in the arms of no less than Claude Levi-Strauss.

His move from physics student to geography student to anthropology professor is radical enough, as was his trans-Atlantic migration to pursue a chosen career. But his innovative move to do lengthy fieldwork among natives, on another continent far from his birth place, was singularly remarkable for the time. Though Bronislaw Malinowksi's stay in the Trobriand Islands of New Guinea during World War I is usually viewed – for reasons I have never completely understood - as the beginning of ethnographic fieldwork involving lengthy and intense interaction while living with natives, Boas' stay in Baffin Land with the Inuit pre-dated Malinowski's sojourn by three decades.

Then the remainder of his career was largely spent working with Native Americans, in the field, in the museum, in the classroom.

His researches among the North West Coast Indians, especially the Kwakiutl of Vancouver Island and environs, is what makes his ethnography famous for most who know of him. This canon of research publication – his over 700 articles and "five-foot shelf" of Kwakiutl monographs (Harris 1968:252, 315) - produced more ethnographic output than many modern departments do collectively in a scholarly generation.

Though he is not famous for developing any named theory (such as Malinowski is for functionalism or Levi-Strauss for structuralism) he is renowned as the father of American anthropology, the founder of statistical physical anthropology, the advocate of long term residential ethnographic research involvement with people willing to act as the anthropologist's teachers, the founder of professional studies of American Indian languages, the man who brought folklore into the academy as a serious branch of study, and a champion for human equality as opponent of racism and supposed hereditary superiority of a given race. As both scientist and advocate he is said to have done more for the cause against racism than anyone anywhere ever. Such was his influence that even though he had not lived or worked in Germany for over forty years the Nazis of the 1930s burned his books and rescinded his Kiel University doctoral degree.

For the first thirty or forty years of the twentieth century, as anthropology grew as a professional discipline, his ex-students, then their students, occupied leading academic positions in almost all major departments across the United States. His major scientific paper "The Limitations of the Comparative Method in Anthropology" (Boas 1973 - the original was published in 1896 in *Science* 103) almost single handedly laid the foundations for the destruction of the prevailing paradigm now known as nineteenth century evolutionism – whereby people and their cultures had been shoehorned into stages,

Savagery, Barbarism and Civilisation, passing in unilinear fashion from one to the next as they progressed or advanced or evolved.

This re-focus of what the disciplinary pursuit should be about, resulted in the obligation on anthropologists – or at least North American ones interested in language and society – to spend time in the field collecting data from natives, trying to understand their point of view of the world. Which in turn led to a much broader view of others, in the way they were influenced by their respective histories and cultures. Culture became *the* key concept championed by Boas for understanding others. They were not barbarians, occupying some inferior stage of evolution, but simply different people, shaped by different histories and cultures. So-called armchair anthropology – practised by men in places like London and Paris pontificating their theories, who had never, in some cases, seen or spoken to a native - became a thing of the past, a mode of speculation demolished, due to Boas in America, and also to Malinowski in Europe.

Thus was launched a Kuhnian paradigm. Forget the claims that Boas never gave us a major theory. He gave us a way to practise our profession, a philosophy of practice, if you like. One that was based on real people, or their biological and archaeological ancestors, their histories, and their cultures. One that gave due respect to different groups of people, seen as subscribing to their respective belief systems, communicating in their respective languages. And one that where possible used large number samples to provide insights from what would become statistical anthropology.

As an aside, scholars like Marvin Harris (1968) have devoted much energy to deriding the lack of theory in Boas' work. They claim that because of this he has been accorded far too high a place in the anthropological pantheon. This is not the place for evaluation of the pursuit of such claims, which needs far greater volume and multi-faceted response. I will simply say these things: 1) scholars like Harris live in (heavenly) glass houses; they should be careful when embracing theory as the only mode of worth to be pursued by top shelf researchers, that their own simplistic functionalist take on theory is actually deserving of anything more than stone throwing; 2) to pour scorn on someone like Boas whose work bristles with reference to theories and hypotheses, while laying down massive amounts of data, of empirical evidence, with which to assess those theories, is to pour scorn on a class of empirical researchers such as Lord Rutherford, such as Marie Curie, such as the genius Michael Faraday, such as Galileo Galilei. These men and women, some of the greatest scientists who ever lived, not only never wrote a theory, but were in some cases - Rutherford's for example - downright scornful of what he dismissed as those "damned hieroglyphics" that constituted the science of mathematical physics.

In what now seems a profoundly simple and maybe even uninteresting breakthrough, Boas showed the scientific world and the world of learning, that people and their cultures were merely empirically different. The so-called stages model was an invention of nothing more than European arrogance and prejudice. But when Boas wrote and published his greatest paper, things were not that simple or uninteresting. He had a fight on his hands to get social scientists to understand that their work had to be opinion-free and evidence-based. Data, not ideologically driven speculation, were needed. It was not scientific to merely sit as a believing christian in gentlemen's clubs in London and pontificate in an Indo-European language about the lowly status one perceived – through one's gin and tonic - of folks from other cultures who spoke other languages and subscribed to other belief systems. That was radical.

As I said above Boas also, virtually single handedly too, created North American anthropology. Columbia became the seat of the discipline, with an entire generation of famous professors trained there as students of his (think: Alfred Kroeber, Robert Lowie, Elsie Clews Parsons, Ruth Benedict, Leslie Spier, Melville Herskovits, the genius Edward Sapir, Ruth Bunzel, Margaret Mead, Alexander Lesser, Paul Radin, Alexander Goldenweiser, E.Adamson Hoebbel, Ashley Montagu, and many more). That was all pretty radical as well.

Then way ahead of his time, and out of step within it, came the attempt by Boas "to strengthen American Negro identity by an affirmation of the value of African culture" (Stocking 1974b; see also Boas 1974c). It has certainly prospered and assisted profound social and historical implications and consequences, many of which are ongoing. I doubt Muhammad Ali ever heard of Franz Boas. But the old man could sting like a bee when it came to confronting prejudice with scientific evidence. As the Nazi book burnings attest, he knew how to deploy data that got under the skin. This aspect of Boas' use of science in advocacy has been judged by George Stocking as no less than "radically innovative" (Stocking 1974b; see also Boas 1974c).

Searching for cultural influence, from the remains and artefacts of earliest hominids through to the way contemporary speech acts are modified and maintained among living native speakers, became the lynchpin of Boasian anthropology. Thus did the English speaking world of academic anthropology divide, between the culture based interpretations of the Americans, and the more sociological approach followed, and still pursued, in Britain. Through this major innovation, Boas created what became known as four field anthropology (which I have noted in my Introductory Section).

This Section however, is intended to deal with none of that, really, other than to introduce you to the great man and his greatness in anthropology. Rather, later in this Section I am going to discuss another very radical thing Boas did, bringing to his discipline methods and techniques from the realm of what we might now call statistical anthropology. In George Stocking's (1974d:59) evaluation, "Boas' statistical orientation ... in the period prior to 1911 was probably as sophisticated as that of any scientist in America." It is indeed deeply radical that such a distinction would lay with an anthropologist in any era, much less back in those pre-SPSS times.

But first some background related to the title of this Section. How does statistical anthropology fit with the fundamental laws I refer to in title? And, what are these great laws?

Well, the fundamental laws of science which are instrumental in giving us the keys to a statistical anthropological approach are essentially grouped threefold: (i) the Second Law of thermodynamics, including Boltzmann's Law; (ii) Shannon's Laws of information theory; and (iii) the laws underlying frequency distributions, especially the Gaussian or normal curve, and its moments.

Let me provide a brief discussion of these. First, the Second Law of thermodynamics.

Lauded by Peter Atkins (2010:xii) as "one of the all-time great laws of science," as it "illuminates why anything ... happens at all", and said by Arthur Eddington to hold "the supreme position among the laws of Nature" (quoted by Shannon & Weaver 1998:12), it follows work initially done by the Frenchman Sadi Carnot on what is now known as the Carnot cycle. Then Rudolf Clausius and William Thomson (later Lord Kelvin) independently gave us the empirical basis of the Second Law of thermodynamics. Fermi (1956:30-31) provides us with both versions of the law:

"A transformation whose only final result is to transfer heat from a body at a given temperature to a body at a higher temperature is impossible" [Postulate of Clausius];

"A transformation whose only final result is to transform into work heat extracted from a source which is at the same temperature throughout is impossible" [Postulate of Lord Kelvin].

In other words, the Second Law empirically establishes the fact that heat cannot flow spontaneously from a colder body to a hotter one, nor can it do work while its source remains at constant temperature.

The Zeroth Law of thermodynamics gives us the notion of temperature; the First Law the notion of energy and its conservation (Atkins 2010; Fermi 1956). The Second Law introduced the concept of entropy. This can be thought of as a measure of change in any amounts of heat that flow. Entropy is also thought of as a fundamental thermodynamic property that characterizes the quality of energy (Atkins 2010). Increases in entropy mean irreversible changes in a system, making it possible, for example, to explain time as unidirectional, as an arrow, the metaphor most used by physicists and chemists – and, historians. Those irreversible changes occur in time, or over time, giving us the arrow metaphor for time. Because entropy is a one way street, as it were, there is no going back; thus we cannot reverse the flow of time. The Tardus and the Terminator rely on the notion of travelling *backwards* in time, a very different idea from the reversal of time.

Physically and mathematically, entropy is defined as the quantity of heat per unit temperature change in any such situation. It is a global concept, where an isolated system can be anything from a laboratory container to the universe itself. Entropy can decrease locally in more complex isolated systems such as these, but globally it can never decrease; it must always stay constant or increase. The modern way of stating the Second Law briefly and elegantly is thus: in the face of spontaneous change, the entropy of an isolated system never decreases (Fermi 1956; Atkins 2010).

Entropy, as introduced and discovered by Clausius and Thomson was an empirical phenomenon, as I said. But it was the great Austrian Ludwig Boltzmann who added for science a theoretical perspective on how entropy might be considered. Boltzmann was a theoretician seriously committed to an atomistic view of matter. In fact his greatest work is widely considered to be the development of statistical mechanics (in which the properties of matter are seen to derive from the behaviour of atoms). He copped much flak for this from many of his contemporaries, eventually leading to his suicide, aged only 62. But courtesy of developments that came decades later in the work of Gibbs and Einstein, his ideas prevailed. Historical redemption was his. In fact, for those of us reared and educated in the West after World War II, it now, with hindsight, seems ludicrous that scientists could have ever held views opposed to Boltzmann's, opposed to what we now know to be the way things always are. But, that's history; and it's the way things sometimes are.

Boltzmann's atomistic view of matter and nature led him to interpret entropy as the probability of states to take certain configurations. He derived what is known as Boltzmann's Law, which states simply that entropy is a function of the probability of states to take certain configurations. In fact, more formally, entropy is proportional to the logarithm of the probability (Fermi 1956:57). That is, the function resolves into an equation – which is the form of the law he proved – through that probability being factored as its logarithm times what is known as the Boltzmann constant. This famous and elegant formula now graces not only every textbook in thermodynamics, but also Boltzmann's tombstone in the *Zentralfriedhof* in Vienna.

The Boltzmann Law, the theoretical take on the Second Law of thermodynamics, can be better understood, perhaps, via notions of order and disorder. Imagine, for example, two containers connected by a valve. In one container there is gas, in the other nothing - it is evacuated. We open the valve. Gas flows from one container into the other, spreading itself out fairly evenly throughout the two containers. We refer to the initial conditions as highly ordered: the gas is contained and settled in one situation, one configuration. Opening the valve we arrive at a situation of greater disorder, the atoms of gas have migrated through the two containers. Disorder has increased. Boltzmann's genius was in seeing this as how entropy works: disorder has increased; which is to say entropy has increased.

If that example is too reminiscent of the horrors of high school science to be enjoyable, think of another example in which two equal sized adjacent paddocks are joined by a closed gate. One paddock contains kangaroos, the other: empty. The gate is opened. Now, leaving aside the realities of living animals to herd together, to merely stay put or follow each other into the second paddock, theoretically, the kangaroos will spread out, some remaining in the first paddock, others investigating the second. The density of kangaroos, in terms of say, animals per square metre of the first paddock, has decreased, their distribution has become more evenly spread or dispersed between the two paddocks. The paddock and kangaroo setup has become more disordered. Entropy of this isolated system has increased.

Another and final example concerning the same two paddocks. At the outset one paddock contains kangaroos as before. But this time the other paddock, rather than being empty, contains wallabies. The gate is opened. Curious investigative individuals from both species move into the adjoining paddock. The result is that our two paddock system is now fairly evenly populated by a mix of kangaroos and wallabies. The system has again become more disordered; entropy has increased.

Below we will see another way of thinking about this: each paddock contained only one species; highly ordered paddocks. But with gate opened, the disorder of two intermingled species has increased. We say the diversity of species in any given paddock has increased.

Yet another way of thinking about the system is as follows. The spontaneous changes – or transformations - that followed valve or gate opening, take the respective systems to a state of higher probability. And furthermore, we find that the most stable state of such a system will be the state of highest probability (Fermi 1957:57).

So increased entropy signifies increase in disorder, which signifies in turn increase in probability, which yet again signifies increase in diversity. This sequence leads us along the path from some of the most significant laws in physical science to our investigations in statistical anthropology.

Along that path I turn next to a brief discussion of Shannon's Laws in information theory.

Claude Shannon was one of the greatest scientists of the twentieth century, and arguably the greatest American-born scientist ever. He virtually single handedly created what is now known as information theory, which underlies what we know globally as IT, information technology. He was truly "a hero of the information age, a man without whose razor-sharp insight none of this technology would work" (Aleksander 2003:214). Yet so few ordinary people recognise the name of the man who gave us the foundations that allow all of us, in the affluent world anyhow, to have our computers, laptops, tablets, techno chips, hard drives, computer networks, mobile phones, fibre optic cables, entertainment devices, and so on. Both a mathematician and an engineer, he "brought these disciplines together in a way that changed the world forever" (Aleksander 2003:214).

There are two laws due to Shannon that we need to take note of (see Shannon & Weaver 1998). The first states that the amount of information required to gain knowledge of an event is proportional to the probability of that event occurring (do you hear the echoes of Boltzmann?). The less likely an event is, the more information its occurrence carries (Aleksander 2003:221). Or, in other words, "information is proportional to how much you don't know" (Aleksander 2003:221).

The second Shannon law tells us that the amount of information that can be transmitted through a line depends on two factors, the bandwidth (or range of frequencies) and the signal to noise ratio. Aleksander (2003:214) gives a nice example for understanding this second law. Imagine yourself in conversation in a crowded noisy pub, such that you have to shout to be heard (increase signal to noise ratio). Imagine also you are talking to a partially deaf person (whose bandwidth is therefore restricted), you have to shout even louder. More formally, this equation does two significant jobs at once: determines the rate of information transmission, and gives a figure for the amount of information that can be stored.

Shannon brilliantly adapted the Boltzmann Law to link information to probability, or as we may see, diversity. In fact it was the restatement of that law that gave us the first index of diversity, such as adopted in ecology (Magurran 2004; Walters 2013). His second law adds another dimension to this diversity aspect, giving insight into the number of species or classes or categories we might expect within this information (Macarthur 1975; Walters 2013).

I turn now to the third of my groups of great laws: those concerned with normal frequency distributions, their character, and their allies.

Every undergraduate student in the sciences or engineering, in economics, psychology and sociology, is familiar with the normal curve of first year statistics. Some learn to write down the formula for such a distribution, or in some cases, even derive it. In my experience as an anthropology student then teacher in Australian universities, that claim doesn't necessarily hold for members of my discipline. In my era most Australian anthropology students would have graduated uncontaminated by a course in introductory statistics. Sad, that. There would have been a few students of biological anthropology who would have become acquainted with such material, and perhaps even the odd archaeology student as well. But not many. And most of those students would have gained their familiarity through use of a department-purchased copy of the Statistical Package for Social Sciences (SPSS). So perhaps anyone reading this who was trained in that sort of anthropology may first need to become cursorily acquainted with an introductory text in statistics and probability.

The normal curve, or Gaussian distribution, named after the wonderful Carl Friedrich Gauss – another great German genius – who did much of the mathematics related to what I am about to briefly describe, is a frequency distribution most common to the way random variables assemble.

What does that mean? Let me explain by example. Suppose we want to get an idea of the height of female netballers. But we have neither funds nor time to try to track down every player in the world. So we have to take a sample. We restrict ourselves to say, Melbourne, and we obtain player lists from the relevant clubs. From these lists we draw a random sample of say, 100 girls. With their agreement to take part, we then meet these people and measure their height. We then draw a simple x, y graph, with heights on the former axis and frequency, or number of occurrences, on the latter axis. We pump our measurements into the computer which then produces a frequency distribution. This will most likely be in the form of a bell shaped curve, with few measurements at either end, while the bulk of measurements increasingly sit in the centre region of the graph. This is our normal curve.

All additive frequency distributions of this sort are – to use the jargon – asymptotically Gaussian. This means that as we increase the size of our sample – by measuring more netballers – our graph will eventually come to take the form of a normal curve. Small samples can be a bit wonky, due to various factors, but we can be confident that as the sample size increases, the frequency distribution will take this shape.

That's all very good. But what do we want to do with this information? Well, we might be doing a survey for a clothing manufacturer, and are keen to see how height of netballers compares with height of same age females in the general population. So we also sample 100 girls from other walks of life, sales girls, wait staff, teachers, tennis players, checkout chicks, for example (ensuring none of the above are also netballers). We get a frequency distribution for them too. Then we want our computer to compare one sample against the other. To do this we use some quantitative measures related to normal curves. We call these Moments of the normal curve. They are mathematically defined by calculus applied to probability, and all have formulae for calculation, mostly in the notation of fairly simple algebra, but I won't tease you with those details here. Brief and somewhat rough descriptions will suffice; and anyhow, some of them are well familiar to many readers. Here they are:

The Zeroth Moment of the normal curve simply gives us a probability of certainty – that is, the probability equals 1 – that our measurements will lie under our curve.

The First Moment is a measure of what is called central tendency in the distribution; that is, it gives us the population mean, (or median or mode) which we approximate by the sample mean (or median or mode).

The Second Moment gives us a measure of dispersion in our data, given as the population variance, or its square root, the standard deviation, which we can apply to our sample.

The Third Moment describes the way in which – with our sample still insufficiently large – the curve varies away from normal; that is, it might have a longish tail projecting out one end of our graph. This we call skewness, and we say such a curve is skewed to either right or left.

The Fourth Moment describes how the peak of our frequency distribution is either "thin" or "fat"; that is, is it a blobby peak to the curve, or a sharp, thin peak. This measure we call kurtosis. Obviously a normal curve will have what we call simply, a normal peak. Kurtotic peaks will either be thinner or thicker than this.

There are other Moments, used by statisticians for various theoretical ends, but essentially for we mere data users and interpreters, the above are sufficient for our purposes. In fact, most data interpreters is research disciplines like anthropology will rarely use moments beyond mean and standard deviation. These are usually enough for us to perform a simple test to detect differences between samples. For example a simple test involving means and standard deviations of our two samples might show, say, that female Melbourne netballers are, on average, taller than our general sample of girls from other pursuits, and their heights are not as varied as are the general populous. The means and standard deviations give us – and our clothing manufacturer – an idea of just how tall, on average, they are, what the distribution of their heights looks like, and which group might be taller than others.

Combinations or ratios or products of our moments can also be deployed to describe various aspects of frequency distributions. One most commonly used is the coefficient of variation, a simple ratio of standard deviation to the mean. Another example is the Gini coefficient (Section 9) used by economists to describe inequality in frequency distributions of wealth. It is essentially a measure of the dispersion or spread of a distribution.

Statistical thinking also leads us into the area of probability. How probable are certain outcomes? What are the chances that certain events will occur? Significant developments in the statistical arena allow us to assess such chances, to make judgments about how likely particular outcomes are. A key example of such statistical and probabilistic thinking involves making decisions under uncertainty with the help of a discovery known as Bayes' Theorem. We can use Bayesian analysis, or Bayesian inference as it is usually called, to strengthen our ability to conclude or interpret from data sets and samples.

Now, in our netballers example we might have chosen to analyse the data not by simply plotting each measurement on the graph, but by grouping measurements into classes or categories. Let's say we grouped our height measurements into classes like 180-189cm, 190-199cm, 200-209cm, etc. Then we could have had our computer allocate each individual height measurement to one of these classes, then perform our normal curve plotting as before, only this time it would be the frequency of netballer heights in each of the chosen classes that gets plotted by the computer.

This form of analysis has been with us since the first developments of statistical thermodynamics. For example, Schrödinger (1989:2) advises: determine the distribution or "state-of-the-assembly"; classify them; count the numbers in the classes; judge the probability of certain features or characteristics of the assembly.

Such a method proves extremely useful in dealing with measurements or other data that fall naturally or by design into classes. I am thinking, for examples, of such data as the number of creatures in particular species, such as birds on an island, or insects in a paddock. Another example might involve numbers of archaeological sites in particular type classes or site types, such as walled towns, encampments, stone scatters, refuse middens in an area chosen for archaeological survey. Or again from archaeology, numbers of temples, offices, ball courts, houses, in the remains of an ancient city. From ethnology, we might like to analyse the number of individuals in particular tribes or bands or segments or groups, such as clans or lineages. Or for linguistic anthropology, the number of languages or dialects spoken by individuals in any or all of these last mentioned social groupings.

For this kind of analysis we find that once again as we add numbers to make our sample sizes large, these assemblies also take the form of normal curves. Two special forms of this occur as what are called logseries distributions (usually for smaller samples) or lognormal distributions, (for large samples). Again these derive from the same simple rule as before, that additive statistical distributions are asymptotically Gaussian, or normal. These two distributions are powerfully useful in the analysis and consideration of data that come in classes, types, or species. The clearest and most comprehensive reference I am aware of on these is the classic paper by Lord May of Oxford (1975).

That's enough. These brief and rather sketchy descriptions will suffice to give you a feel for the great laws that allow us to make genuinely scientific claims about samples we choose to draw in anthropological research including, via the Boasian four field model of our discipline, ethnology or ethnography, archaeology, biological anthropology and anthropological linguistics. However, we must never forget the old adage: garbage in, garbage out. As always, our science will only be as good as our assumptions, our logic and our evidence-hugging interpretations.

I have said all I need to say about laws as background. But let me sum up what is immediately implied by the relation between these great laws and their usefulness to a statistical anthropology. What that, and hopefully this book, teach is that the power we gain from statistical anthropology comes from three fronts of attack:

1) Where possible find, deploy and use large numbers.

Make your samples as big as you possibly can. This is not always easy. But if you are forced to work with small samples plan to do comparisons or collations so data can be amalgamated, tiered, combined. Large numbers don't always lead in straightforward ways to the correct answer. But they sure help.

Of course studying the materials our profession researches will always mean that encounters with small samples are commonplace. But in such cases we need to be aware of the statistical techniques which allow us to make valid scientific inferences nevertheless. I intend no holier than thou claims here, for my own work has meant dealing many times with small samples (e.g. Walters 1987, 1996, 2000, 2013). But it is important to confront such samples with methods and techniques involving procedures for dealing with the dynamics of frequency distributions, in particular the fundamentals of the normal curve and its familiars, as well as other measures and analyses that can bring some degree of sophistication to our inferences.

2) Know and use fundamental frequency distributions.

Have ready at hand the key frequency distributions relevant to your questions, your methods and your data. Know them and know the mechanics of how such distributions assemble. Be familiar with what these processes mean for your analysis. Knowledge of frequency distributions and how they assemble will mostly place your conclusions far closer to the truth than will comparisons of mere descriptive numbers such as moments of the normal curve.

3) Use quantitative indicators.

Nevertheless be prepared to use the moments of the normal curve, whether that be mean and standard deviation many are familiar with, or various others. But much more importantly be prepared to investigate data patterns further using measures of diversity, disorder or probability to give indications of how frequency distributions compare and contrast as well as how with increased confidence decisions can be made under uncertainty..

This is the essence of statistical anthropology. The remainder of this book provides some examples of these principles at work. So let us begin. To do that, in the following section I will return first to my beginning, and the man who started it all: Franz Boas.

Three
Boas & Statistical Anthropology

Early in the twentieth century, ensconced by then at Columbia University, Boas took advantage of the thousands who were sailing, fleeing the harsh realities of Europe, heading across the Atlantic for what they saw as the land of opportunity. He worked with the United States Department of Immigration to study groups of immigrant arrivals from seven identifiable ethnic groups, sampling "a total of almost 18,000 persons" (Stocking 1968:177) or, as it is reported in Table 18 of his monumental tome, 17,821 people (Boas 1912:84). His measurements were physical, bodily sizes and shapes, the presence or absence of features. Boas asked: are morphological characters set in stone, as it were, by hereditary history? Or are they malleable, changeable according to environmental factors and conditions?

Thus did the great man initiate the first professional research work in statistical anthropology. However, Boas often acknowledged the scholarly debt he owed to Francis Galton and Karl Pearson in developing "methods of the quantitative study of the varieties of man" (Boas 1974a:32).

He was able to do follow-up studies on these immigrant samples, so that he could track growth and development in the New World against their initial state upon arrival. Of course, children grew, but adults changed too. Boas was able to lay cause to such changes to their new environments, in terms of living conditions, hygiene, diet, and so on.

There have been some academic papers from the decades late in the twentieth century suggesting repudiation of Boas' findings. But those studies have themselves been questioned, and in one key case well repudiated. The critical re-studies of Boas' data have mostly been shown to have flaws in beginning assumptions as well as conclusions and interpretations, leading to them being convincingly discredited. In fact, one author was able to show the data from one such study, when cleansed of the aforementioned assumptions, actually supported Boas' conclusions (Allen 1989).

Boas was good at planning for large samples and getting data on them. For example a study he was entrusted with on behalf of the Department of Ethnology of the World's Columbian Exposition involved a physical examination of North American Indians. The material Boas was able to put together involved measurements from "about 17,000" Native Americans (Boas 1974e:192). In another example that harks forward to Levi-Strauss, he even undertook a "quasi-statistical study" of the distribution of folk-tale elements (Stocking 1974a:13).

But that landmark immigrant study - and the other large number ones - while they established statistical anthropology, are not my focus in this book. The argie-bargie of academic to-ing and fro-ing on those environmental findings will possibly never be settled. (Although it has gone quiet over the past two decades.) Some people will probably still not accept that Boas made ground-breaking advances in this area that contradicted much of race theory. Other researchers, finding strength in his methods, data and conclusions, will not be daunted by these gainsayers.

However, I will leave that issue for elsewhere. It is to another related study of human diversity that our attention needs to be turned here.

Boas was born into and grew up in a Germany imbued with racism and racial stereotypes, aided and abetted by a scientific establishment, in this case mainly medically trained anatomists, who identified the formalism of hereditary characteristics of morphology. To these researchers and their political benefactors and teachers, such attributes were inherited, held fast during lifetimes, and were useful in identifying races and their consequent place in a stage centred model of human evolution. People were graded and classified as higher or lower according to the sum total of their measured morphological characters.

This had significant follow-on consequences for oppression and disenfranchisement. Policy based upon bigotry and ideology was seen to be justified by science. Jewish people, such as Boas and his family, were one group who suffered at the hands of this politicised and confected research. Ideology and everyday prejudice took the place of genuinely scientific questions, hypotheses and objective testing. This misuse of science had an unfortunately long pedigree in Germany and other European countries, reaching its apogee under the Nazis in 1930s and 1940s Germany and its conquered countries. In addition to the awful genocide we are all familiar with, it is disturbing to read of the extent to which Hitler bastardised science, closing down much pioneering and cutting edge scholarship to replace it with – in universities and research institutes – so-called departments of Race Science which taught race theory and race superiority but included as well technical departments teaching methods which were to be deployed in order to facilitate and accompany genocide (Moore 1989:340).

Presumably such techniques included things like removing gold from teeth of corpses, hair for pillow and mattress stuffing, operation of gas chambers, crematorium ovens, etc. These last would see many technical-level graduates employed in the death camps tasked with procedures and outcomes which no civilised society would surely ever allow to happen.

In typically radical fashion Boas railed against all this, seeing that science required the asking of questions, followed by testing of hypotheses through data collection, before emerging with interpretations and conclusions. No place was there for ideology and bigotry. Neither was there place for preconceived prejudices. It is possible that all this shaped Boas' working methods, the ideologically based theories being repugnant and in need of evidence based assessment (hypothesis testing). This in turn gave him the scientific approach he deployed in practice throughout his career.

Certain later claims seek to play down Boas' large scientific influence. Some of these are based around the notion that Boas was not theoretical, that he had no theory and bequeathed us no theory (Harris 1968:250-318). But it was not that he had no time for theory. His writings show him constantly addressing claims of theory. He used empirically garnered evidence to test those theories, to assess them as viable or no, and if the latter, to reject them. Just like Rutherford and his "boys" (Farmelo 2009) – his graduate students - were busy doing in the Cavendish at about the same time.

For our purposes here, concerned about statistical anthropology, we see Boas, from his earliest times in America, thinking about observed information through "actual distributions of empirical phenomena" (Stocking 1974a:3). This was particularly evident in his physical anthropology where he opposed attempts to subsume the "range of distribution of measured physical differences within an idealized type" (Stocking 1974a:3). Such types which were defined in terms of average values had to be re-defined "in terms of the total distribution of variants" (Stocking 1974a:3). An apparently "normal" distribution may be the result of many kinds of complicating factors, such as an example he gave of the "intermixture of two different groups" (Stocking 1974a:3) that he found among the Great Lakes Indians.

We see Boas as a researcher of large samples. He was an analyst who placed emphasis on distributions and was wary of curves that appeared "normal" but may have hidden deeper meanings. Thus was he wary of simplistically deployed moments that might lead to false conclusions about distributions, such as complex assemblages being reduced to ideal types.

New to the Americas, one of the first debates Boas became involved in concerned cause and effect in the context of evolutionary ethnology. It began from the standpoint taken by the prominent anthropologist Otis Mason that like causes produce like effects. Boas came out strongly opposed to this, arguing that like effects do not necessarily have like causes (Stocking 1974a:2). Implicit in this was the issue of classification, which was the focus of much of his early work. The attack on Mason, according to Stocking (1974a:3), was centred on Mason's attempt to define "families, genera, and species" of ethnological phenomena that could then be treated comparatively. Involved in this was the notion of "like effects." Boas criticised this from the position that we had no prior claim to judging the likeness of effects. Museum exhibits and displays brought it to a head. Boas attacked these "technological species" as "rigid abstractions" (Stocking 1974a:4). In ethnology, he claimed, "all is individuality" (Stocking 1974a:4).

Boas has been vindicated over time, and his opponents (for the argument involved far more than just Mason) and their approach consigned to history. But I will however, take this opportunity to revive this issue in a starkly tangential manner. It will involve taking a stance that at first appears to contradict Boas' position. But a closer look will reveal that my method is very different and does not involve anything to do with rigid abstraction (see Section 8). I will argue here and show in Section 8 that provided we don't go near unwarranted evolutionary implications, and provided we are not involved in issues related to cause and effect, we can make use of a very modified classificatory scheme that does not involve rigid abstraction from individuals, to assist our interpretations in some areas that lend themselves to such a method.

Taking inference from the fundamental laws I concluded the previous Section by suggesting that we should try wherever we can for large samples. While taking this stance I acknowledged this cannot always be done. But there are ways that we can think tangentially to give some effect to our samples that allows them to behave as if they were large. I suggest that one way to do this is to classify them as species, genera and even families. Then we can not only compare, but we can employ quantitative measures such as diversity in ways that are independent of sample size. I will hold off on this now, knowing you can read more about the method and argument in Section 8.

Four

Frequency Distributions of Archaeological Sites: Sumeria

Conflict in the Middle East does more than ruin potentially good geopolitics and destroy many innocent lives. It has collateral damage attached as well, on less visible fronts; to archaeology for example. Not only are current campaigns of violence destroying world cultural heritage treasures, they prevent enlightening research programs from taking place in peaceful, welcoming contexts.

I take as example the significant American-led research programs in Iran, which happened in the 1980s, led by distinguished researchers Gregory Johnson, Henry Wright and Robert Wenke (Johnson 1987; Wright 1987; Wenke 1999). Among other things they compared urbanisation and city states in various regions where different state entities arose in the mid-Holocene.

Brief background: they examined a period of early urbanisation; research uncovered a "series of emergences" of "individual states in a network of politics" (Wenke 1999:408). By 5200 years ago conflict between the elites of Susa and those of Chuga Mish reflects an attempt by the latter to sever its dependent relationship on the former.

Between 5,000 BP and 4350 BP Sumeria had some 13 city-states (Wenke 1999:412).

So what does all this have to do with statistical anthropology? Well, one of the key features they examined in their surveys was the size of the archaeological sites they visited and recorded. From these raw data they were able to have their computer construct frequency distributions. For Sumeria, Gregory Johnson used rank-size distribution to demonstrate a pattern for settlement size. This kind of distribution involves ranking sites by size before the program graphs and analyses them.

Johnson found many smaller sized settlements with smaller and smaller numbers of ever larger towns and cities. Now, this is where statistical anthropology proves useful, in helping with interpretations of situations such as this. However, not everybody sees the same interpretations. For example, in his summary Wenke (1999:409) claimed, "no one really knows what the different kinds of rank-size plots mean." For they "are essentially empirical generalizations without any theory to explain them" (Wenke 1999:409).

Really? Statistical anthropology comes to our aid here. The distribution Johnson detected reflects hierarchy, with settlement size showing the same kind of pattern as distributions of individual wealth, occupations, and other indicators of Very Big Hierarchy (VBH) societies (Walters 2015). For example, in Sumeria we note the following hierarchical elements: god-king, nobles, wealthy businessmen, scribes, artisans, smiths, slaves (Wenke 1999:419). For other VBH contexts such as ancient China: king (with supernatural powers), hierarchically arranged nobility, commoners.

As far as theory goes, beyond the key role for hierarchy (Walters

2015), the mathematics of statistical theory and our great laws accounts for such distributions in a straightforward manner. As I discussed in section 2, our prevailing law comes to the fore: once frequency distributions grow large and assemble the way VBH variables do, we expect (and presumably Johnson found) the distributions to be log normal.

Had he, and especially Wenke, realised this simple interpretation from some very basic mathematics to do with the way frequency distributions involving data in classes, assemble, the interpretive conclusion follows in a straightforward manner.

Statistical anthropology teaches us that the simplest and most logical conclusions are not always cultural or social – or archaeological. They are sometimes mathematical, and follow from some very basic, clear and powerful laws.

Five

Catch Curves in Archaeology

It is the 1980s. Australian archaeology is caught up in its most quantitative phase ever. The effects of Binfordian functionalism are everywhere felt. English functionalism is present as well. At this time many Britons, as well as British-trained Australians, occupy staff positions in Australian departments. Marvin Harris' cultural materialism – another version of functionalism - has also attained some following, especially where there is American influence. Young archaeologists, especially graduate students, are counting everything they can excavate, find, collect, or seek out in museum cupboards. Especially is this true for stone artefact analysis and the analysis of faunal remains.

I am enrolled as a PhD candidate at University of Queensland. This is a department strong in four field anthropology, the Boasian model. My topic (Walters 1987) is the analysis of fish remains excavated from Holocene middens around the shores of Moreton Bay, the large embayment that opens the way to Brisbane, the state capital.

Prior to entering anthropology I had initially completed a science degree in the Zoology Department, having specialised in ecology. Evolution and taxonomy also had significant presences in my program. I felt confident in handling animal remains for my study. I was also well versed in statistics and probability, as well as some other areas of mathematics.

At that time the predominant class of site types among the recorded coastal archaeological sites on the continent would have been the class of shell middens. So a generation of us became involved in gaining expertise in molluscan shell morphology, taxonomy, age and growth, as well as being able to identify and analyse vertebrate remains associated with these. In my case, with the Moreton Bay middens, there were precious few remains other than shells and fish bones. I left the molluscs to others, focussing on remains dominated by vast numbers of vertebrae, with smaller amounts of other bone fragments plus a few otoliths.

I did not lack for literature. A healthy set of graduate students had preceded me, in some cases by a few years, while one or two academics had also entered the fray. A few papers had been published, but most of the data were to be found in Honours theses from various universities. In those days these were fairly easily acquired. Most of our supervisors were acquainted with one another, so sharing of such unpublished literature presented little problem.

In addition, this branch of the discipline had also taken off overseas, giving us a strong English-language literature from New Zealand, the UK and North America. I garnered as much of this as I could. Most of it was methodological, something that was unsurprising in a relatively new area of academic study. Focus came to be on how we identified various body parts, and what species they derived from. Then the more interesting interpretations followed: what sizes the animals had been, the age structure of the catches, the methods that might have been deployed to catch them. It was all heady stuff.

As you will have realised by now, these last mentioned questions implied dealing with frequency distributions, while also using statistical methods to make various inferences. Statistical anthropology to the fore. Below I will briefly set out one example from those times whose message remains strong and clear for us today.

In the 1970s and early 1980s Australian archaeologists (Allen 1972; Kefous 1977; Coleman 1978, 1980; Blackwell 1980; Bowdler & Lourandos 1982; Balme 1983; Sullivan 1984) introduced and developed the idea that non-recoverable remains of prehistoric fishing implements could possibly be inferred from the catch data of fish bones they excavated from their archaeological sites. In doing this they worked in tandem with developments, as I said, that were going on in New Zealand, North America and Europe (e.g. Casteel 1972, 1974, 1976; Leach 1979; Colley 1983).

Reconstructions of fishing gear were then attempted by focussing on two lines of evidence: i) the types and diversity of fish that specific kinds of gear could be expected to catch; and ii) the attributes of samples of fish remains recovered from archaeological sites.

In relation to the first point, because of various species-specific habitat distributions and behaviours, certain forms of gear are more amenable to catching certain kinds of animals. For example, fast moving upper pelagic predators are unlikely to be found in very shallow waters of mangrove associated mud banks, so will not lend themselves to being taken by hand nets held by catchers standing in water near a mangrove bank.

Secondly, key numerical attributes of some kinds of gear are related to key numerical attributes of animal size. For example, a large hook is useless to try to engage in the mouth of a small predatory fish, so the hook size has to be correlated in some way with the bite size of the prey being sought.

So, the reasoning went, if we could identify the types of fish found in the sites, then knowledge from the literature on behaviour and habitats of those species would tell us something about the kinds of gear that catchers would need to deploy to have a chance of catching them. Collaterally, it would also provide prehistoric habitat and environment information allowing reconstruction of what the area under investigation was like back then. For example, if the fish sample came from a species of fast moving upper pelagic predators the archaeologists could be fairly sure that the local habitat was fairly deep open water. And if said species readily took bait on the hook, provided the size of the hook was within a range that fit the gape size of the species, it might be inferred that hooks of such and such a size were deployed in the hunt.

This all constituted a very clever piece of reasoning and invention. But would it work in practice?

The archaeologists involved were convinced it would and it did. As the studies came along and the publications flowed it seemed a whole new code had been cracked in the investigation of material culture lost to sight by methods of normal recovery of objects from the past. Here we could, it seemed, determine the gear the prehistoric fishers had used without ever having access to it. It mattered no more that net fibres didn't preserve, or that shell or bone fish hooks decayed away in the soil formation processes of the aeons. The fish bones would tell us their story.

And a good story it was too. Until a certain party-pooper (Walters 1987) came along, during the work he undertook on the quantitative study of another set of fish remains, to show that some of the assumptions used by the archaeologists had been a tad over optimistic, granting more certainty than was warranted to inferential arguments that followed. I turn to these now. This issue which may seem of little interest nowadays given the way that quantitative studies seem to have declined in archaeological popularity in places like Australia, remains one of historical import. It provides for us an insight into the uses of statistical anthropology, locating its centrality in interpretations that involve appreciation of both the full range of empirical evidence and the underlying mathematical workings of such processes.

So to the assumptions. The archaeologists relied on such maxims as: animals can often be identified from their remains, be they of carapace, shells, teeth, bones. It turned out to be fairly easy to identify many kinds of fish from key skeletal elements preserved in sites. It was also well known that the size of fish was related in some way to the size of the skeletal elements that gave structure to their bodies. There are complications with this sometimes, but in general big bone implies big fish, small bone small fish. These two key assumptions were and are true.

However, when the archaeologists deployed other assumptions in order to take things further along the chain of reason they began to step onto shaky ground.

Firstly the archaeologists tended to assume that there would be some kind of straightforward one-to-one relationship between on the one hand kind of gear used and types of fish caught, and on the other hand attributes of that gear (such as mesh size of nets) and the size distributions of fish caught. It did not take much investigation to turn up an array of studies from both fisheries biology and ethnography demonstrating that while this was true very generally, any relations revealed were neither straightforward nor of a one-to-one type correspondence.

The studies from industrialised fisheries and from ethnographic records of living fisher folk showed many particular fish species being taken by various methods. For example a lagoon species might be captured on various occasions by spearing, by netting or by use of poison. All went in the same pots, and all bones went on the same middens, regardless.

In regard to size returns again the assumptions did not hold up at any more than a very superficial level. Yes, there was a very general relation between size of gear and the frequency distribution of caught sizes, but these relations varied vastly. In fact in the studies I utilised the relation of net mesh size to fish length varied by a factor which ranged between 10% and 40%. This was far too broad and general to fit the tight conclusions the archaeologists wanted to draw about things like the mesh size of the nets they postulated the prehistoric fishers were using.

They also made a vague assumption related to some relation between the sizes of fish caught and the structure of the catch distributions. They added to the confusion by using vague terms like "normal populations" where there was possibly an intent to imply normal distributions. Data and theory from fisheries biology showed that this too was supported somewhat vaguely (aside from the "normal" confusion), but that again care had to be taken with inferences.

It is easy to take a quick look at a table of five size classes, say, represented in an archaeological sample of fish remains, see a few in the smallest class, a lot in the next, big numbers in the third, then less and finally small numbers again, and call this normal. Fair enough. But when you have claimed, as they were doing, that certain kinds of net produce catches which take the form of normal curves, you must be more attentive to technical detail. Allowing me one rather pedantic point using numbers here from my argument back then: good fisheries theory showed that while many such catch curves were in fact normal (the archaeologists were correct) many were skewed away from normal. The Russian theoretician Baranov had derived from the fisheries data a relation that showed the net type in question produced catches in which only 2-3% differed from the optimum (median or mean) by 20% (quoted in Hamley 1975:1944; Pope et al 1975:5). All the Australian data differed from it by amounts that ranged between 6% and 27%. The curves were not tightly normal and the inference of certain net with certain size was totally misguided.

A second nail in the metaphorical coffin came from introductory statistics. To show this I beg indulgence one last time for the appearance of some pedantic numbers. Normal distributions are characterised by having two standard deviations either side of the mean cover some 99.7% of the distribution. The Australian data which were supposed to show strong normality in the catches and thereby once again support inferences about the use of a particular kind of net being used comprised percentages that covered two standard deviations as ranging between 57% and 90%. What emerged clearly was that the samples were not strongly normal, but were in fact the opposite: not tight, and quite skewed. On a second front the inference of certain net with certain size was again totally misguided.

Enough of the petty nuts and bolts of the archaeological samples. While they were important to a technical argument what was and is of far greater importance is an examination of the underlying mathematics that tells us how both hunting factors and archaeological site formation processes work to produce the results that excavators and analysts find in their data.

Thankfully that is extremely straightforward too.

The "normal" curves of the fish bones, which weren't all that Gaussian at all, turned out to not be well accounted for by the one particular kind of gear of one particular arithmetical attribute which had supposedly been prehistorically deployed to accumulate them in their shell midden deposits. Like most assemblages of archaeological fish remains they were far more likely the result of many different catches deposited over a considerable time span using a variety of techniques and items of gear. Size frequency distributions that build up in this manner over such times are described by what is known as the Central Limit Theorem of statistics (Lapin 1973:224; May 1975:89).

This is a theorem which sits at the very heart of statistical theory, the one that tells us so richly about the attributes of the *statistical* normal curve – as opposed to the usual or "normal" curves of the archae-ologists. In a nutshell it is a theorem to the effect that "all additive statistical distributions are asymptotically Gaussian" (May 1975:89). Which is to say that if hunters keep throwing their fish remains on a midden heap, over time the frequency distribution of bone size will approach closer and closer, as the sample size increases, to a normal distribution.

As such archaeological assemblages come to represent the additive accumulations of multiple samples – reflecting multiple fish catches – the composite catch curves come to represent normal curves. In addition the statistics of these combined samples will provide bet-ter and better indicators, as sample size increases, of the population parameters of the available fish. Where by available, fisheries scientists mean sizes and species that are amenable to being caught using any given gear. Which is to say, the archaeological samples have the potential to describe, perhaps above some critical lower size limit, the size parameters of fish in that local habitat, irrespective of the gear used to land them.

The upshot of all this was to point out that instead of waving hypothetical smarts about that flagged some particular kind of gear with particular size attributes being used to catch the fish concerned, we would do better to follow the empirical lead given by both industrialised fisheries data and ethnographic fishing data from living fisher folk to emphasise a simple maxim: same technique, using same gear, in different ecological situations results in different catches; different techniques, using different types of gear, in the same ecological situation results in similar catch curves.

The pattern of species abundance in catches was not to be seen as due to some simple functional relationship of taxon to item of gear. Nor was the pattern of size frequency distribution so due. Knowledge of prevailing ecological or habitat conditions at times in the relevant past loom large in importance for understanding the patterns of fish bone occurrences in excavated shell middens.

And in order to achieve this it is so important to have at hand some understanding and appreciation of the statistical features of importance. The patterns in the bones are mathematical patterns, and the mathematics should not be ignored as a crucial explanatory factor in accounting for empirical research outcomes.

Back then I advocated (Walters 1987:167-171) the notion that some quantitative measures might helpfully be deployed to characterise such assemblages, rather than just calling by eye that they looked "normal." Zoology and botany are, and were loaded with such quantitative measures (Magurrin 2004) that could be used to good effect. I also suggested a much closer mapping of fisheries theory and data together with archaeological data would be handy, to refine some of the factors that merely blurrily linked gear attributes and catch curves. I saw good archaeological analysis being a big help here to fisheries theory. This never happened.

I also noted that assemblages constituting the archaeological catch curves conform to the so-called "hollow curves" known formally as logseries or log normal distributions. Patterns of abundance in such samples show some degree of numerical domination by one or a few taxa. In order to convincingly demonstrate some direct relation between kind of gear used, or indeed size of some key attribute of gear, and the animals caught more had to be done than merely find some vaguely related "normal" curves in the samples. The use of some quantitative measure of diversity loomed large as the best possible answer. This never happened either.

In addition the examples I presented in Walters (1987) showed that size frequency distributions of animals caught tended to normal or skewed away from normal curves for all kinds of gear. Ethnography showed that contemporary natives fishing with non-industrial techniques in the Pacific and elsewhere frequently used multiple types of gear to land the same fish within the time frame in which an archaeological midden of the discarded remains would have been deposited. The resulting frequency distributions of the sizes of fish caught reflected not gear used, but fish ecology, the habitats in which they were taken, the mathematical effects of multiple additive sampling, and archaeological site formation processes.

STATISTICAL ANTHROPOLOGY

It remains a shame that this potentially useful and important set of innovations that arose in that era in archaeological faunal studies seems to have died with the fall out of fashion of all such quantitative archaeological analysis. Perhaps a future generation of analysts will return to the science some day, this time bringing more sophisticated statistical anthropological techniques and considerations with them.

Six

Catastrophic Culture Change as Transformations in Statistical Distributions

Consider a hypothetical terminal Pleistocene cum beginning Holocene Domestic Mode of Production (DMP) society (Sahlins 1974; Walters 2015) which can be characterized by a set of what I will term state variables. As the society changes into what archaeologists and anthropologists have termed a complex society (or in antiquity referred to as civilization), or as I have termed it a Very Big Hierarchy (VBH) society, the state variables take on a different character or different individual "values." Some intensify into new forms. Others appear which are radically novel and emergent.

Let us further consider the ability to document these changes via an independent variable which is social and cultural. This would allow us to characterise these societies, the old and the new, as a function of some key independent social or cultural variable. Shortly I will identify this independent variable, but first let us examine the changeover from DMP to VBH societies in terms of the frequency distributions of wealth and possessions available to citizens in a hypothetical example of each kind of society.

In a typical DMP society we find the distribution of wealth and possessions to follow what is called a uniform distribution. That is, each member of the society, given the usual anthropological caveats of age and gender, has about equal access to the resources available to the society at large. That is, the amount of wealth available to any individual is given by the total wealth available divided by the total number of individuals in the society.

The switch away from DMP to VBH society gives us a social form whose wealth distribution is vastly different from uniform. It follows what is termed a log series distribution. This is a distribution where a few of the individuals hold most of the wealth, another class or other classes of individuals hold middling to smaller amounts of wealth, and the majority of individuals have little wealth and few means. This is a distribution for societies with relatively small numbers of individuals (as archaeologists tell us the initial VBH societies would have been). But it is akin to the log normal distribution that describes wealth distribution in large societies such as those of modern global capitalism (see Section 9).

In this hypothetical VBH society we see that the wealth available to any individual is given by an algebraic expression which relates that amount of wealth to the fraction of total wealth appropriated by the highest ranked individual initially, the fraction of the remaining wealth that is then taken by the second ranked individual, the fraction for the third highest ranked and so on (May 1975). I will call this relation the appropriation parameter.

The transition from DMP to VBH society is marked by key factors in these distributions. We see the DMP society characterised by a wealth distribution that goes as a function of one divided by the total number of individuals involved; in other words with approximately egalitarian distribution of resources throughout the society. Whereas the VBH society has the appropriation parameter, which characterizes its wealth distribution. In other words, wealth concentrated in one individual, with a lesser amount appropriated across the distribution until the great bulk of individuals are seen to have not very big shares at all.

Some other points we can draw from this relation are: the change in distributions from uniform to appropriated carry with them the implicit invocation that society is increasing is size, in population numbers, from one to the other; as society increases in population size, it draws proportionally greater resources to a comparatively smaller part of the population spectrum; and finally this last also suggests that more resources are being or are capable of being harvested or harnessed by the society. But those are essentially side issues for the transition, consequential effects rather than anything causal.

The transition from one society to the other is marked by a shift from one to the other key parameter (uniform to appropriation). The dimension of both these can be considered as equivalent to some fraction of wealth (which we can assume to be linked with some measure of power and prestige) per person. When this is scanned across the entire probability distribution we have what can be seen intuitively as a measure of hierarchy. Given the above, the variable I mentioned earlier can now be considered as hierarchy. In Walters (2015:75-80) I developed an index for hierarchy to account for its role as a variable of consequence in this transition.

So how did it happen? The discussion of change or intensification comes to centre upon key social and political players recorded for DMP societies, figures known variously as Chiefs, Divine Chiefs, Potlatch Chiefs, Big Men, etc. We can characterise what might have happened if for example a Big Man (BM) drove the onset of VBH societies by marshalling surpluses but refusing to redistribute them as happened in the past.

A statement from Sahlins (1974:87) sets the context.

> The three elements of the DMP so far identified – small labor force differentiated essentially by sex, simple technology, and finite production objectives – are systematically interrelated. Not only is each in reciprocal bond with the others, but each by its own modesty of scale is adapted to the nature of the others. Let any one of these elements show an unusual inclination to develop, it meets from the others the increasing resistance of an incompatibility. The normal systematic resolution of this tension is restoration of the status quo ('negative feedback'). Only in the event of an historic conjuncture of additional and external contradictions ('overdetermination') would the crisis pass over into destruction and transformation.

Thus do we witness six million years of DMP social organisation. These are societies conservative in the limit, while paradoxically being culturally dynamic, determined to retain what is was they possessed, something one might romantically regard as a quality of life never known by man or woman since. Negative feedback was their great ally. Contradictions their enemies and seeds of destruction. And it is these that begin in ever so small steps to intensification, where the "success of only a few" becomes an "invitation to violence" (Sahlins 1974:88).

Then it happens. As the structure is politicised, especially as it is centralised in ruling chiefs, the household economy is mobilised in a larger cause (Sahlins 1974:130). For political life is a "stimulus to production" (Sahlins 1974:135). For example, the systems of status competition found in Melanesia develop economic impact from the "ambition of aspiring big-men" (Sahlins 1974:135). Leadership "generates domestic surplus" (Sahlins 1974:140).

Harris (1979) identifies Big Men as nodes to intensify production. The crucial step for him is the rise of chiefdoms, which he links to the appearance of the BM. Influential older men whose advice and guidance the community seeks and follows, are found at the centre of events. They not only intensify production through urging, cajoling, brow-beating, pleading, they also carry out redistribution of centrally collected goods, and lead the way in fighting and trading. They will come to an office the society is now willing to have established.

But after a long pedigree of DMP redistribution, more and more of the centrally collected surplus remains with the BM, as he begins to alter the balance between redistribution and appropriation. So before, in the DMP, where we saw a uniform distribution of wealth, this is now being replaced by the rapidly changing social structures in which the BM garners it all, redistributing less than all, then only a fraction.

Before we know it, the BM has become a Rogue Individual (RI), a BM gone mad – mad in the sense of overthrowing a history of tradition, morality and political behaviour, against all apparent DMP logic. Of course he can only do this if he can convince followers that his is the true way forward, that the office he now wishes to occupy is necessary for social success, that the new symbols incorporated must replace the old irrelevant ones to bring benefits to everybody.

The RI appropriates, with support from followers if need be, and force where necessary. Beyond his court, he now sees no need to redistribute. These are the new rules, within a new social structure, which have to quickly be made into laws and embedded in new approving – or at least fate accepting - belief systems. The new frequency distribution of wealth across the transitioning society is a geometric series, the log-series referred to above. After the RI appropriates his slice, his princelings follow with theirs, the new generals with theirs, and so it goes on down the line. Most people now merely get bad news. And the planet gets the catastrophic change to massively hierarchical VBH society.

Notice that the key to understanding this issue involved two key components, one being the central nodes of social organisation and political power: the BM and the RI. The other being the statistical distributions that mark for us the frequency of wealth available to members of the societies, and the changes that were marked by the changing distributions.

Notes to Section 6:

The onset of VBH society can be described as catastrophic in two senses. Firstly in the technical, mathematical sense of catastrophe theory (Thom 1989; Poston & Stewart 1996; Arnold 2004). As I wrote in Walters (2015) the change can be described theoretically using continuous models, but it also invites use of the perhaps more relevant and possibly more sophisticated catastrophe models.

Secondly, and this is more judgmental, I suggest that on a number of indicators the change to VBH societies can be seen as having long term catastrophic consequences for our species, for a large number of people alive and dead who have had to live through poverty, squalor, destruction of their homes and lands, and finally for the planet itself.

Seven

The Social Organisation
of Australian Tribes

Joseph Birdsell was an American biological anthropologist who spent most of his career researching the tribes of Indigenous Australia, their physical morphology, ecology and evolution. In work spanning the 1940s through 1970s, much of it in league with the great mapper Norman Tindale, he provided many insights into Australian social organisation.

It was Tindale by the way who produced the first comprehensive map of the tribal distribution of native Australians at and around the time of European contact. This map was still in vogue through the 1970s and 1980s, after which I presume more detailed contemporary land claim and Native Title work gave new intricacies and nuances to knowledge of land ownership, language distribution and social organisation. However, one can still readily find on the internet modern versions – attributed to various folk and organisations – which are essentially renditions of Tindale's monumental work.

The lead study of concern to me here however is one that demonstrated a correlation between tribal area and environmental productivity as marked by rainfall patterns. Birdsell collected data on tribal areas defined linguistically or at least in regard to dialects spoken, mapped (with Tindale's expertise) for size and shape, allocated Indigenous names for all tribes concerned, and obtained local albeit modern rainfall patterns.

Birdsell's graphs of tribal area against rainfall were beautiful to see, data points located on the two-axis space, and a line of best fit drawn in. This was one of the so-called "hollow curves" of an inverse correlation: the areas with lowest rainfall, covering much of inland Australia showed large tribal areas while small areas were packed into coastal locations especially in the well watered tropical north and the sub-tropical and temperate eastern seaboard.

Sampling hundreds of tribes gave a robust fit to the data. When grouped in terms of numbers of tribes with various areas, we saw again nice correlation between a large number of small tribal areas, lesser numbers of medium sized areas, and far less frequent numbers of large to very large tribal areas – mainly in inland and desert locations. Thus the data gave good fit to what we know from statistical anthropology as a log normal frequency distribution that I have discussed in earlier sections.

Making excellent use of statistical anthropology Birdsell gave great insight into the social organisation of Australian tribes in particular relation to resource productivity of the environments in which people made a living during the period of European takeover.

When it comes to the social organisation of Australian tribes the name we think of immediately is Radcliffe-Brown. The famous English ethnologist published as three parts in the journal *Oceania* what together became a key monograph on the topic (Radcliffe-Brown 1931) – whose title I purloin for this section - which for decades remained a key document in Australian studies. Perhaps it still does. Here I will consider a tribe according to the definition provided by Radcliffe-Brown (1931:5). A tribe consists of persons speaking one language, or dialects of one language. Its unity is primarily linguistic. The name of the tribe and the name of the language are normally the same. There is also a unity of custom throughout the tribe.

Each tribe may be thought of as occupying a territory. But this is because it is composed of a number of "hordes" each of which has its own territory. Hordes in turn, which in modern parlance are called bands, are the major local group. So the territory of a tribe becomes the sum of the territories of its component hordes or bands. These consist of a small group of persons owning a territory, which has known boundaries, and over which they hold proprietary rights.

Even though we often and perhaps modally think of ethnography and social anthropology (at least in that golden era 1880-1960) as a lone pursuit with one fieldworker living with and closely studying information provided by a key informant or perhaps a few of them, together with the immediate family or band or local group associated with such a nodal host or hosts. In the modern era, or post-modern era, ethnographers have eschewed generalisation, science and its possibility of laws or law-like regularities, as well as bothering with comparative studies. But Radcliffe-Brown, while he participated in ethnographic field work, lived and researched in an era when ethnologists sought generalisations, drew comparative conclusions, even developed laws, or advocated such development. We have seen this already with regard to Franz Boas.

Radcliffe-Brown (1979:87) let us know his approach.

> If you will take the time to study two or three hun-
> dred kinship systems from all parts of the world you
> will be impressed, I think, by the great diversity they
> exhibit. But you will also be impressed by the way
> in which some particular feature, such as an Omaha
> type of terminology, reappears in scattered and wide-
> ly spread regions. To reduce this diversity to some
> sort of order is the task of analysis, and by its means
> we can, I believe, find, beneath the diversities, a lim-
> ited number of general principles.

We see here advocacy of large samples, used comparatively, involving the key concept of diversity, in the search for underlying
principles. That's science, as practised by the ethnologists who
lived and worked during the first half of the twentieth century.
And when used like this with samples of the form scholars like
Radcliffe-Brown advocated, studies avail themselves of the power
of statistical anthropology.

Indigenous Australian social organisation is structured such that
individual societies or tribes as I have been calling them in the
nomenclature of those days comprise a number of structural elements, which the ethnologists termed "classes." Some tribes have
but one structural concept and term for the entire social group; that
is, one class. Some have two, the tribe being divided "vertically," as
Radcliffe-Brown suggested into moieties. Some have four divisions,
called by Radcliffe-Brown sections, while the final cluster have eight
classes, called by his term sub-sections.

In various small pockets of the coastal periphery of the Australian continent were and are found the tribes that had only the one class. They were known by one term, often related to locality, which distinguished their tribe from adjacent societies. The distribution of these groups is "infrequent, discontinuous, and marginal" (Service 1971:127). See Figure 1 below for a map of this distribution. Yet they exist in the continent's most productive and densely settled areas, the coastal regions.

More widely distributed and more frequent are the tribes bearing two classes. Taken "vertically" these are Radcliffe-Brown's moieties. Two exogamous groups are formed, a person's membership of her moiety being taken either from father's half in some tribes or from his mother's in other tribes. On the other hand, the two classes may be defined "horizontally," which is to say by alternating generations. Father's generation is separate from child's, as therefore are their classes. The two class system was found spread over much of the southern part of the continent, inland Victoria, inland New South Wales, half of South Australia, the southwest of Western Australia, then in various pockets around the Gulf of Carpentaria, Top End and the Kimberley.

Figure 1. Map of social class distribution in Australia. The numerals refer to areas which have respectively 1,2,4 or 8 classes, the last three known as moieties, sections and sub-sections. From Service (1971:129) adapted from Lawrence (1937:348, Map A), drawn by Edwin Ferdon Jr.

When tribes incorporated both the above distinctions, that between generations F and C, together with that between moieties M and N, the society is divided into four classes FM, FN, CM, CN. This system had the widest distribution across the continent, being found from coastal northern New South Wales and Queensland, all the way across to the central Western Australian coast. Almost all of Queensland, most of northern New South Wales and big geographic chunks of both the Northern Territory and Western Australia revealed this dominant form of social organisation, the section system. It was the most common form of Australian Indigenous social organisation at the time of European usurpation.

A smaller number of societies had what are called "semi-moieties" in which each moiety was divided according to proscribed first cross cousin marriage against prescribed second cross cousin marriage. This extra axis of division led to the eight classes of Radcliffe-Brown's sub-section system. The social form is largely found in the Northern Territory and the contiguous part of northeastern Western Australia. While more common than the one class system, it involved a smaller number of tribes than either the two or four class systems.

Now, in the twenty first century must we find this nothing more than antiquated scientism? Are we not allowed to find it fascinating that the inverse relationship Birdsell discovered between environmental productivity (as measured by rainfall) and tribal size can then be mapped reasonably neatly onto a relationship between the class systems of social organisation and once again, environmental productivity. Hence follows, albeit more loosely, the syllogistic relationship between tribal size and form of social organisation.

As Service (1971:127-30) reminded us, the early scholars appear to have been on the money by following the logical inference from these statistical findings. What they saw, inferentially, was a reconstructable history of the class system, an evolutionary scheme if you will.

D.S.Davidson, one of these early scholars, wrote in the 1920s that the most "logically permissible conclusion" (Service 1971:128) was that initially all tribes had a one class system. Two classes appeared across part of the continent, perhaps as people expanded into new areas, then similarly four classes followed and finally eight. The one class system was eventually left where they had been since the beginning, on and around the productive coast. Except that now with two class systems in vogue, the one class system was left in isolated discontinuous pockets. The eight class system was a relatively recent innovation according to this scheme, budding off from the four class system when some tribes adopted the idea that instead of marriage to cross cousins generally, there should be a prohibition on first cross cousin marriage while second cross cousin marriage became prescribed. The two numerically and geographically dominant forms, two classes and four classes, remained the most widespread over the continental mass once drier areas of the inland were occupied permanently.

Service (1971:128) also reminds us that there are two kinds of two class system: named exogamous moieties and named adjacent generations. As Radcliffe-Brown had suggested the four class system came about when tribes saw advantage in combining these two organisational schemes. Why, he asks, are the compound systems later and central (Service 1971:128)?

It is helpful to see each system as dependent on a previous simpler system. As the map shows, one and two class systems occur in coastal areas where population density tended to be highest, tribal size smallest, and where people did not have to travel extensively or distantly in search of food. The arid regions of least population density and largest tribal geographical size saw local groups having to travel widely for food and water. This is where the four class system originated and spread. Here it was most common, with the eight class system budding off from it relatively recently.

A.P.Elkin, another of the early scholars, made the observation that in the drier areas of the continent foraging groups must be relatively small, and be widely separated, often for significant lengths of time. Hence it becomes more comprehensible why the compound systems tend to be located in arid regions (cited and quoted by Service 1971:128-9). The eight class system, for example, spreads at the expense of simpler ones because of its greater utility in intertribal relations. One system can be used by all tribes meeting and greeting. If each of the language groups had its own special system only confusion would result. The invention of the more complex system is thus seen as something of a cooperative venture between groups (Service 1971:130).

Compound class systems become useful in the interactions between people whose meetings are sporadic and rare. These are people who most likely will not be familiar with the finer and more complex status characteristics of each other but who do need to know the most significant and gross characteristics of each other in order to initiate at least ephemeral social relationships. The early ethnographers described it exactly thus: a system that worked well in intertribal gatherings.

As Service (1971:131) also acknowledges, these ideas are due largely to researchers like Radcliffe-Brown, Davidson and Elkin, ethnologists who were able to see the broader canvas, to comprehend the large in the small, appreciate comparative studies, seek generalisations, and like Birdsell who would come a little later, to understand and apply brilliantly their observations using the bigger statistical picture.

Eight

The Species Concept and Sex Work Research

Diversity, as pointed out in the Introductory Section, was largely developed originally in communications and information engineering and technology, then feverishly expanded in biological ecology. It has also found its modern uses in physics, chemistry, palaeontology, archaeology and museum studies. Recall that essentially it is simply a method of quantitatively examining an assemblage of categories (bits of information, species of animals, types of bones or artefacts) to see how varied or otherwise that assemblage is. In that sense it is a measure of how even the assemblage is with respect to its component categories, hereinafter referred to as species.

Imagine an assemblage of 10 species in a given space, each with 10 individuals. We would say this grouping is at maximum diversity, with all species even in terms of their numerical contribution to the makeup of the community. Another assemblage of 10 species, but with one species having 60 individuals, another 20, yet another 15 and the remainder each having 5 individuals, would show a numerical dominance by one species (60 individuals), others less so (20 and

15), rather than an even spread. This group or assemblage would have a lower value of diversity than the first community.

In Walters (2013) with the view to addressing certain important questions about variability and resilience in sex industries, I introduced a diversity measure into a study of sex work. Notice from the above paragraph that diversity most commonly involves a combination of the number of species with the number of individuals in those species. However there are profound difficulties in attempting to quantify numbers of prostitutes in these situations. Nevertheless both Hershatter (1997) and I – for different stated reasons – saw value to be had in, or inferred from, the attempts. Because numerical values of species are pretty much ethically and bureaucratically impossible to glean with accuracy and precision, it becomes useful to find a way to consider diversity without worrying about numbers of individuals concerned. Fortunately there is an easy and ethical way. For the first, most convenient and most simple measure of diversity is a straightforward species count.

In this Section I focus on those species, what they are and how they can be deployed. In other words I introduce a species concept. Such a notion is well known from biology, but I claim it can also useful to sex work research. This is a modified species concept, one vastly different from the biological species original, and constructed according to different criteria. Given that, I will describe a concept that is surprisingly flexible yet sufficiently rigid to be quantitatively useful. As I showed in Walters (2013) such quantitative usefulness then leads directly in certain cases to testable theory. From the resultant theory a case can be made that potentially brings important practical benefits for sex work and sex workers.

Let me begin by reiterating that there is no delusion here that we can equate such a species concept with the quite firm notion used in biological evolution. What I have in mind is something far more fluid. There are no equivalents to capacity for successful reproduction, a bounded breeding group, the survival of fertile offspring, and so on. This species concept is much more fuzzy. While the biological concept fences individual organisms in to a particular category, the concept I have in mind depends on categories with gated fences, so that individuals can move freely between one category and another. Individuals are born into biological species, live in them, reproduce in them, die in them, and sometimes fossilise in them. But here, there is nothing equivalent to saying individual sex worker I_i belongs in species P_p as a lifelong or career long category. In a given working career a particular sex worker might be a street walker, a brothel worker, a call girl, a girlfriend, a temporary wife, and a karaoke girl. Six species represented in one career (lifetime) as a sex worker or one fraction of that career. These different species can manifest themselves at different stages of a sex work career, one species following the other. Or they can be somewhat contemporaneous, such as a brothel worker who moonlights, once she returns home from her shift at the brothel, as a call girl or escort. Two species in one day (or night).

In considerations of the biological world, species as bounded breeding entities are grouped together according to similarities. That is, according to attributes shared. We humans, for example, carry vast differences physically and mentally, and in personality and behaviour. Yet our shared DNA, reproductive biology and other features allow us to interbreed successfully. That is, to produce viable offspring, as members of the same species. Should Anglo Saxon Lord Highfalutin breed with a Jamaican slavegirl of Afro origins, the result will be a human baby. Should a native American from Tierra del Fuego breed with a Hmong girl from the uplands of Laos, result:

the same. We are all in this together. Each and every one of us are members of the species *Homo sapiens*.

And what is more, our species, like all other biological species, is bounded, fenced in. Should Lord Highfalutin try to breed with his terrier, the results will not only not be a human baby, but will actually produce no outcome at all. That is, we cannot produce viable offspring when attempting breeding with members of other species. Occasional cross-breeds like mules (horse with donkey) or ligers (lion with tiger) do occur, but the offspring they produce are not reproductively viable. They are living dead ends. Because their parental species are bounded reproductively.

At one elementary level this is the case with prostitution. Street walkers are street walkers no matter their locale of work, their age, their targeted clientele, or the language and currency they use in transactions. Bar girls are bar girls the world over. As are brothel girls. Members of each of these categories share attributes with each other that they may not share with members of other categories. Street girls work in streets, bar girls in bars, etc. At such elementary level these species are bounded too. The key differences here are that individual biological species members are, as I said above, born into and die in categories. Individual prostitutes may move during a career, a year, a week or even a day from one category to another, then another, and so on, as circumstances and whim demand. The categories are not fenced, but gated. The individual members are able to switch, to relocate, to alter behaviours. So at all levels above and beyond the most elementary, the kind of species concept I am advocating here for prostitution shows no boundedness.

The idea of a taxonomy lets us group species together into other categories. These groupings of various closely related species are called genera (singular: genus). In similar fashion groups of closely related genera are grouped together into families. For example, biological taxonomists call our domestic cats by the Linnaean species name *Felis catus*. (The Swedish taxonomist Carl Linnaeus was the originator of this system, which he developed in his famous tome of 1735 – and many subsequent editions - titled *Systema Naturae*, The System of Nature.) Domestic cats are closely related to other cats such as the wildcat *Felis silvestris* and the Chinese mountain cat *Felis bieti*. These in turn are all related to other forms such as the lynx, for example the Canadian lynx *Lynx canadensis* and the bobcat *Lynx rufus*. And again all are relatives of the far larger and more ferocious lion *Panthera leo* and tiger *Panthera tigris*. Notice that we have three genera of cats here, each with its own particular species. They in turn are all members of the Family Felidae.

We can usefully deploy such a taxonomic hierarchy such as this in discussions of sex worker categories. In Walters (2013) I suggested various species of sex work could be grouped together in the category (genus) whore, and it in turn could be lumped with other genera into a family level taxon which I termed prostitute (or sex worker). Such a technique allows us to bring into focus certain ideas which prove useful for examination of historical changes in sex work profiles and services.

My argument's point of departure was an example where I discussed the notion of temporary wife. This was a category introduced into the prostitution literature by historians of the Early Modern period in Southeast Asia (Reid 1993a, 1993b; Andaya 1998). In Walters (2013) I presented seven case studies from contemporary Vietnam which show the continuity of this phenomenon into present times.

What do we mean by a temporary wife? This is a girl or woman who meets up with a foreign man who arrives to do business for a period, moves in under the same roof with him, providing him sexual services and some degree of domestic care, possibly even assisting him in his business, or if she is experienced enough, with interpretation and translation. In return the foreigner provides for her during their liaison, and rewards her financially, with either money or in earlier historical times before money came into use, goods and gifts and materiel. The historians (Reid 1993a, 1993b; Andaya 1998) saw it as a category distinct from the category prostitute, whence they described the transition between, over historical time, temporary wifedom and prostitution. I built on this (Walters 2013), interpreting temporary wifedom as merely another form of prostitution, not distinct from it. A new theory emerged from this claim which led to a prediction that periods of what I termed whoredom would be found in Early Modern Southeast Asian sex work to occur between each period of time spent as a temporary wife. This prediction, as I said in Walters (2013) is testable by historical research.

My development from this key notion of the historians depended upon viewing the situation differently from them. It required thinking in the fluid manner I suggested above for types of sex work. That is, temporary wife and prostitute are not simply different categories *per se* that change from one to the other over historical time. My suggestion was that we see instead one being a subset of the other: temporary wife is one type of prostitute. If we take prostitution as a catch-all category consisting in this first instance of temporary wives, whores, girlfriends (a particular Vietnamese-using-English term for a longish relationship that involves money transactions), street walkers, brothel workers, etc., we have begun to develop a taxonomy for sex work. Within such a taxonomy, each category (such as the examples just given) can be seen as species. In such a simple taxonomy, prostitution would be equivalent to a family level category with various genera each potentially containing species made up of individual sex workers. To the most simple and useful of these genera I gave the name whore, a genus which then contained various of these listed species.

It is important to realise two things – unfamiliar to the biological taxonomist, but probably all too familiar to those anthropological ethno-taxonomists who specialise in folk classifications used by many cultures. One, as stated already, is that these categories may be fluid, at least within the self or life history of any given sex worker. The second is that these categories are in no way meant to be all inclusive or collectively exhaustive. That is, others coming from different cultural and social contexts, researchers, analysts, commentators, sex workers themselves, may recognise and name many other species or even some additional genera. I merely keep my example here simple, to cover the main points I wish to emphasise.

The usefulness of the species concept comes in giving us a way to measure the diversity of sex work in any given context or situation. And that in turn proves not only useful, but essential to the development of productive theory and the predictions we can make from it. This is not about labels; it is not about one sex worker joking with another that she is a call girl, while you – she says to her friend – are a what? Reply: an effing street walker, you slut! No, it's not about that. For each given sex worker, as I say again for emphasis, can and usually will change species various times in a career or even during any given working day or night. Nobody has to be defined or labelled or pigeon holed. Nobody has to be fenced in. Nobody has to be stereotyped. The categories are the things of interest here for analysis. While biological individuals get on with breeding that makes relevance for taxonomic categories (species, genera, etc.), and hence are the central level of importance for biology, here in sex work research, it is the categories that are central to our concerns. But again, these are species, genera, families. Hence the field analyst's question would be something like: what species of sex work are carried out here in this city/place/state/country? It is a broad categorical question. It is never intended to be a specific question addressed to individual sex workers.

Let me elaborate my claim for usefulness of the species concept by providing an example from the histories of sex work in China and Vietnam. These histories relate to the communist government take-overs of both countries in the 1950s and 1970s respectively, periods before those takeovers and the modern period since the late 1980s. Data from these times can help us assess proclaimed governmental success in eradicating prostitution.

The Chinese Communist Party came to power in 1949. They made and continued for decades to make the claim that by the 1950s they had eradicated prostitution (let's say arbitrarily 1955). Prostitution was seen a part of the fabric of decadent governments that ruled prior to the takeover by the communists. Yet by the late 1980s or so prostitution was thriving and growing throughout the land. What had happened? Global and Western morality were seen to provide the supply source of various causal factors that turned Chinese people back to prostitution. That the Communist government was unable to slow, stem or prevent these developments does nothing to lessen their boasting about their efficient elimination of it decades earlier.

The communists in Vietnam make a claim, echoed by government institute researchers and media, that after 1975 they too eradicated prostitution. So why in the late 1990s and early 2000s were various researchers including me able to do productive studies on sex work in Vietnam? Because the government claims that prostitution began again around the mid to late 1980s. Why? It returned, they claim, as a result of social changes following the new policy known in English as renovation, which took place in and around 1986. Markets were opened up, foreign investors began to be welcomed, and vast amounts of international money began to cascade into the economy. The government claims these foreign influences were instrumental in bringing back prostitution. From then, it grew rapidly and diversified over time. That they were unable to slow, much less stem or prevent these developments does nothing to lessen their boasting about their efficient elimination of it just over a decade earlier.

In Walters (2013) I undertook an examination of prostitution in the three historical periods mentioned above. For China (i) the earlier pre-communist period dates as prior to 1950; (ii) the communist removal period when prostitution is supposed to have been eradicated extends from 1955 – 1988; and (iii) the present or modern period, that is, roughly the period after 1988 through the 2000s. For Vietnam this means (i) the earlier pre-communist period before 1975; (ii) the communist removal period 1975 – 1986 when prostitution is supposed to have been eradicated; and (iii) is the same as China.

Gail Hershatter (1997), the central protagonist in this research on China, asked whether the claims of the Chinese authorities were actually supported by the historical evidence. In response, here are some examples from a literature that shows clearly what the Chinese were claiming was without objective foundation.

For 1980s China Hershatter says that foreign reporters were commenting on the "increasingly visible" prostitution occurring in hotels and coffee shops they frequented. Exactly when during the decade is however difficult to assess, for Hershatter cites five relevant articles, including two from 1988 and 1989, but two others published in 1985. I will opt for a conservative approach in assigning them to a period. These comments I judge to be applicable to both the communist removal period (1985 publication dates) and the modern period 1988 and beyond (1988, 1989, 1992 dates).

Political refugees from Guangdong gave accounts which were published in Hong Kong in the late 1960s and 1970s. A former teacher told how at that time two forms of prostitution had been practised. Hostesses in government guest houses, where foreigners friendly to China were entertained. The hostesses were middle school graduates trained in English. Prostitution was also occurring in drama troupes, where daughters of landlords and government officials made money sleeping with "influential party personnel." Though these reports were unverifiable, they gave a picture of prostitution as a

> direct consequence of government policies, a sign that China was badly governed by hypocritical officials who claimed to have eliminated prostitution even as they enjoyed prostitutes themselves and procured them for foreign friends.

> (Hershatter 1997:332)

One group of educated youth who had been held in detention centres, apparently for political offences, reported encountering prostitutes there. Most of the sex providers were young urban women whose parents had been sent to the countryside during the Cultural Revolution. These "adolescent daughters" were left with no one to support or supervise them. In detention they began sex work.

One report of a married rural woman providing sex to village men in return for "salted fish and meat" brings echoes of what it might have been like in the Early Modern period before the onset of money economies. How might the prostitutes of that time have been recompensed? Perhaps this communist removal period example provides a hint.

Prostitutes in downtown Guangzhou solicited customers through the use of code words. As the woman approached a man she would ask if he had "a window facing south," an expression referring to connections with Hong Kong and Macao. She was actually inquiring about his access to a convenient place to go with her and have sex.

As Hershatter (1997:332) sums it up: "these stories delineate a gap between official claims and actual social practice." Hence she is led to ask: did prostitution ever disappear in China under the communists?

Micollier (2004:5) provides an insightful rejoinder when she says that "all the repressive measures which were designed to suppress prostitution never worked, although it became quite invisible for some time. Although this may have been only through a tightly controlled propaganda."

Harriet Evans (1997:175) says that according to official figures, between 1986 and 1990 the number of Chinese prostitutes "increased fourfold over the previous five years." That means, whatever the numbers were for 1986-1990, there were a quarter of them in that first period of the 1980s. In addition two reports from authorities told of the establishment of 62 and 68, respectively, prisons, detention and re-education centres for prostitutes in 1986-7. I have made abbreviated reference to this in Table 1 as more evidence for prostitution in that supposed removal period. No further information is given, but even if each detention centre only had a hundred inmates, that's 6,000 prostitutes incarcerated during those years. Which implies how many still active and outside the detention system?

These data for the communist removal period in China are particularly important as they give the lie to party and government claims that prostitution vanished during those 30 years.

In addition the data together with the above comments from the literature make an enquirer ask about Vietnam, and wonder did similar hypocrisy and cover-up occur during the so-called period of communist abolition and removal of prostitution there during the decade after 1975.

Though there has as yet been no similar historical research for Vietnam - that I am aware of anyhow - eventually it will be done. It appears historians will then also expose the myth of completely vanishing sex work there during that vaunted removal period 1975-1986. There is one intriguing hint from a politics researcher: Nguyen-Vo Thu-Huong (2008) argued that while the Vietnamese Communist Party and its bureaucracy continued the flood of sloganeering and moralising campaigns against prostitution, in effect all they could do practically was the occasional round-up of sex workers. Police generated orgies of peripatetic arrests and incarceration of girls in so-called rehabilitation centres where they forced them to undertake menial tasks such as making conical straw hats. (This was supposed to provide them with skills which, upon release, would dissuade them from returning to sex work.)

To address the issue on a second front I extracted from that literature records of species of prostitution present during the so-called period of removal or eradication of prostitution. I compared this with the periods before and after that so-called period of removal. I did the same for Vietnam.

Table 1 lists the species of prostitution gleaned from both historical and popular literature and, for modern Vietnam, the same but augmented by ethnography. The Table shows species of prostitution recorded during the three historical periods. It also gives relevant references for these. The figures in parentheses at the ends of various cells of the Table show the numbers of species for that country and period. For the removal period the claimed zeros of eradication are augmented for China by the number of species recorded by historical research (in parentheses).

Note, as I intimated above, that I do not imply that the species listed are all there were or are; merely that these are the ones I can find mention of in the literature or for modern Vietnam, know from my own ethnography. Some categories are difficult, such as brothel girls, for there may be many kinds of brothels and workers. Same applies to street walkers. The same problem confounds the example of foreign prostitutes where this one category may include the behaviour and habits of migrant Russian Jewesses that may be vastly different in operating methods, clientele and lifestyles from say, Japanese girls working in Shanghai.

Table 1 allows comparison of diversity of types of prostitution for the three historical periods in China and Vietnam as follows. From earlier through modern we see China having species counts 29, 0(7), 24, while Vietnam shows values 14, 0, 29. As I noted the officially quoted figure for the communist removal period in China is zero, but in fact its actual value is at least 7 different types of prostitution (and maybe more if further detailed historical records are discovered). Therefore at first inspection we may be warranted in casting some doubt upon the zero shown for a similar period of communist removal in Vietnam.

In both countries the earlier period (essentially the twentieth century before communist takeover) and the modern period after 1988, show high diversity of types of prostitution. Don't worry about the slight variation in these high diversity values. Theory would suggest that given China's vast area and population, we should see vastly more prostitutes there than in tiny Vietnam. We should also expect more diversity in species or types. But the numbers in Table 1 do not reflect that. I think this is due to two factors.

Table 1. Types (species) of prostitution in each of three time periods for Vietnam and China, with references.

	Vietnam	China	Vietnam References	China References
Modern period	girlfriend, temporary wife, concubine, street stander, street biker, hotel staff, restaurant staff, dancing hall girls, beauty parlour girls, massage parlour girls, cafeteria staff, bus station girls, railway station girls, dyke girls, embankment girls, sea beach girls, bia om girls, coffee om girls, karaoke girls, sexy dancing girls, bar girls, dancing club girls, drink stall girls, movie stars, hairdressers, soup sellers, recyclers, maids, park girls (masturbators), (29)	hairdressers, manicure parlour girls, nightclub girls, dance hall girls, tourist attraction girls, escort service girls, hotel girls, coffee shop girls, movie house girls, theatre girls, bar girls, girls with rented rooms, roadside shop girls, girls in taxis, train station girls, public square and park girls, girls in private houses, guest house girls, karaoke girls, brothel girls, rental arrangements (temporary wives), truckstop hostel girls, cross-border concubines, mistresses (24)	Walters 2013; Le Thi Quy 1993; Jamieson 1993, p.138-9, 316, 334; Sheehan 1990, p.598, 602; Greene 1973; Anon. 1988, p.118-121, 153; O'Keefe 1994, pp.202-3;	Anon. 1998; Hershatter 1997, pp.34-62, 330-347; Williams 1998, p.80, 86;
Communist removal	0	0 (7)*	No historical research data available.	* "few prostitutes" Dutton 1998, p.8; hotel, coffee shop, guest house hostesses, drama troupe girls, adolescent urban whores, village whores, street walkers with code words, Hershatter 1997, pp.330-333; 1980-1985 estimate of scale plus 62-68 detention centres established, Evans 1997, p.175.

Earlier period	brothel girls, bar girls, massage parlour girls, temporary wife, mistress, concubine, schoolgirl, girlfriend, dancing girls, street biker, street walker, truck wash girls ("steam & cream"), restaurant girls, hootchgirls, (14)	brothel girls, singsong house girls, courtesans (shuyu, changsan, yao er, ersan), street walker (pheasants, flowing rafts), tour guides, massage parlour girls, dance hall girls, movie house girls, teahouse girls, restaurant girls, bathhouse girls, beauty shop girls, rooftop garden girls, taxi dancers, vaudeville house girls, vendors (newspapers, cigarettes, fruit), trysting house girls, salt pork shop girls, rickshaw girls, theatre girls, flophouse girls, foreign girls (Russians, Poles, Romanians, Americans, Japanese), seamstresses, servants, striptease girls, fortune tellers, (29)	Walters 2013; Le Thi Quy 1993; Jamieson 1993, p.138-9, 316, 334; Sheehan 1990, p.598, 602; Greene 1973; Anon. 1988, p.118-121, 153; O'Keefe 1994, pp.202-3;	Anon. 1998; Hershatter 1997, pp.34-62, 330-347; Williams 1998, p.80, 86;

One is that I am familiar with the modern Vietnamese situation and was careful to walk the place recording types. The second is that in contrast Gail Hershatter, beyond providing key examples, never set out to list every type of historically recorded prostitution for Shanghai, while other China researchers such as Michael Dutton and Harriet Evans merely dump most categories under the coverall category prostitution. We have to remember that all these cells in Table 1 are samples only, and that has had an effect as well. For as I stressed above, even my list for modern Vietnam is not exhaustive, merely the types I know about at the present time.

As I said in Walters (2013): over to you, historians of Vietnam. The data from Table 1 – where at least seven forms of prostitution are recorded for China under the communists - certainly make that zero in the Vietnam column look lonely and quite likely untenable. Nguyen-Vo Thu-Huong (2002:130) gives the one tantalising hint I have been able to find, that historical research on this may turn up results as I suggest. Following interviews with an NGO social worker and a former member of the Vietnam Women's Union, she concedes that during the communist period of so-called eradication, "some of the trade went underground."

In addition these data show that prostitution is a resilient (after the manner of Holling (2010a, 2010b) and highly diversified family of phenomena. Hershatter (1997) provided a sound suggestion when she said prostitution should not be considered a "unitary occupa-tion." Totalitarian governments in control of law, police, courts, guns, can drive prostitution down if they are so determined. (Hitler and Pol Pot immediately spring to mind.) Some commentators agree that is what happened in both China and Vietnam. But, I argue, to some degree and for short times only. And, related to the resil-ience I referred to above, what is most astounding is that quality research like Hershatter's can reveal the lie in mouthpiece spruiking of claims of extermination, eradication and the like. The data show that for China. First inspection would suggest a similar situation for Vietnam may be on the cards. Historical research on that topic in Vietnam is needed.

From 1988 both powerful parties and governments appear incapable in their inability to control the escape of prostitution from its bonds.

Western or foreign influences, a turn to individualism and the seeking of good times and material gains were all blamed by various authorities and commentators. But on both sides of the Sino-Viet border the great eradicators were rendered helpless to abolish it a second time around.

Finally I turn to the question of what it all may mean. In Walters (2013, 2015) I offered a theory to explain what I think has happened and is happening. The theory addresses what I will refer to as the evolution of prostitution. That is, its change with modification over time and space. This shows that evolution is proportional to the amount of prostitution that any given cultural context will tolerate, and what I call the unstoppability of prostitution. Which is to say, prostitution is to be seen in the theory as an unstoppable idea, or process, or set of events, highly resilient even at high values of impeding or resisting factors. Which is to say that no matter how much abolitionist activity there is, be it moral or physical, prostitution will find a way to continue and eventually flourish again. This last term – unstoppability - is a new characterisation which, as far as I am aware, has never been presented in any literature prior to Walters (2013). But it is highly derivative, as I also showed in Walters (2013, 2015). The underlying theory is based on and linked to a diversity measure, as well as information flow, and in fact can be used to calculate or predict diversity. I deployed it to that end to predict a removal period number of species for Vietnam.

The theory essentially says that while prostitution is not a natural phenomenon, that is, it is not in our genes or in any set of male or female body chemicals, drives or organs, or personality traits for waywardness, laziness, or lewdness, it has become part of cultural and social life, basically irresistible. Which is to say, that once prostitution arose in Very Big Hierarchy societies five thousand years ago at the eastern end of the Mediterranean, it has proved resilient against all attempts to wipe it out, and will do so through to the end. In terms of its tenure on the planet already it makes the terror and abolitionist periods of Hitler, Stalin, Pol Pot and the communists in China and Vietnam, pale into insignificance. Driven down, it has always rebounded. In terms of its future, I repeat: it is unstoppable (Walters (2015).

Using my system of classifying prostitution as species, then examining it over time as I did in Table 1, allows us to suggest several things. One is that, as I have harped repeatedly already, prostitution is highly resilient. Two, is that in any given society of sufficient size and diversity, we can expect prostitution itself to grow and diversify to an upper bound that basically marks the level of tolerance for prostitution in that society. Note here that the upper bound is not set by one party or government, one dictator or one set of death camps and guns, but is worked out at some level of societal complexity, politically, in the everyday give and take, moral intolerance, moral acceptance, the attempts to abolish it, to close it down, resilient efforts to keep it viable, creative tactics to so do. Third - and this augments and adds conviction to my first inspection suggestion above - the theory can be used with the data from Table 1 to predict that the communist removal era in Vietnam – when there are claims for zero prostitution – will actually show at least six types of prostitution prevailing during those years (6.45, or considering uncertainties, say, 6.0 ± 0.5) (Walters 2013, 2015). Again, historical research can test the theory.

To conclude: in this Section I have sought to characterise forms of prostitution through a simple diversity measure. Because Gail Hershatter has done such brilliant research, the like of which, as far as I know, has not been attempted for Vietnam, I called on Chinese twentieth century data for comparison. The figures show that in the early part of the twentieth century prostitution blossomed, with many forms, or species, recorded. Vietnam was similar, though the recording has been done mainly by American war veterans in personal memoirs, and this biases somewhat the taxonomy as the terms used are rarely native ones, and are often little more than pejoratives. In both People's Republic China and the new unified nation of Vietnam, the communist governments drove prostitution down, claiming triumphantly that it had been eradicated. Hershatter shows this to be untrue for China.

I engaged a theory, developed in physics, information theory and biology, from our great and fundamental laws, to show that prostitution could be expected to evolve in such a manner as to overcome such attempts to abolish it. This theory predicts for Vietnam a similar situation to that recorded by historians for China.

The work described in this Section is an example of how arcane theory – mathematical in the original (Walters 2013, 2015) with a scientific sounding central concept – can have potentially profound pragmatic implications and consequences. Sex work can continue as an industry content in the belief that despite the best efforts of gainsaying authorities, it will continue. The pleasant and positive outcomes of the theory make it clear that abolitionists may as well give up and turn their energies and attention to some more useful and positive activities. For in the long haul they will never defeat prostitution.

Sex workers and their support staff should see this as much more than academic wank. The power of this theory is that it gives to sex workers a political confidence never before attained logically and deductively, except in hope and perhaps insightful opinion: that your work can and will continue forever. Yes it will change over time. Yes it will adapt. But no amount of abolitionist activity, lobby group pressure, legislative changes, periods of policing enthusiasm, will stop it. Stand in its way they will, continue to arrest and prosecute they will; make a nuisance of themselves for hard working people they will. But the political fight must be taken up to abolitionists for they have no chance; they will not win. Sex work will be the winner, and in a civilised world of the future, will prevail to the end.

If the theory is recognised and accepted in civilised societies of the future, social power will have been handed to sex workers universally. Then the industry, organised into unions or professional associations, with standard accoutrements of civilised industries such as continuing education, health clinics, child care facilities, etc., for workers, can finally be free of the cheap tawdry organised crime that benefits now and has in the past, from the context prostitution is thrust into: a pariah industry operating at the edges of respectability falling victim to the unruly and corrupt who would prey upon it and its practitioners.

Notes to Section 8:

None of this is to deny, airbrush or ignore deeper underlying realities of a gendered world of female dependency on this form of labour. For example, Stoler (1997) reminds us of the class structures always present, whereby prostitution and concubinage were tolerated for lower class ranks of colonials, as they were seen by the authorities to keep the lid on desire, preventing things from getting out of hand. But there is very much a family dependency as well. The other thing that emerges in my story (Walters 2013) is that this family dependency, as well as family push factors, are decidedly female. My ethnography records mothers and aunties engaged in urging – if not stronger actions – girls to prostitution. There are class and colonial issues that cannot be overlooked. But to play the game I'm playing is hardly to be Orientalist, or to deny colonial realities, or to devalue the political histories that divided and united, at the same time, foreign (not only Western) men and indigenous Asian women.

Regarding these taxonomic categories, would writers like Carole Pateman and Andrea Dworkin include wife as well as approving my choice of terms? Certainly for the latter, in terms of the "metaphysics of male sexual domination" all women are whores (Chapkis 1997:19). However, we need to be mindful when discussing prostitution to realise that it is not an identity – a social or psychological characterisation of any given group of women - but an income generating activity or form of labour (Kempadoo 1998:3). If not, we are left with the problem of the "whore stigma" (Chapkis 1997:103; see also Kempadoo 1998:3). Another issue is that whore has no equivalent in many languages such as Japanese (Scott 1996:15). However, the term whore has a longstanding precedence in common English parlance, its derivation being from Scandinavian languages whose earlier forms were given as *hore* and *hoore* (Scott 1996:15). Scott also says that when the revised version of the Christian New Testament was prepared, harlot was substituted wherever whore had occurred in the older version (Scott 1996:15-16). In more recent times whore has become a self descriptor of choice among many sex workers, whereby some rights activists have "recovered and valorised the term" (Wheatcroft 2001:10). Some examples: the first sex worker based internet discussion forum was self-titled Whorenet (Murray 2001:150); Wheatcroft (2001:55) reports sex workers in Australia referring to themselves as "good whores"; the International Committee for Prostitutes' Rights held a series of World Whores' Congresses from the 1980s (Pateman 1988:200).

Griffin (1999:13) says of taxonomies in another context, this "is what men have done to women, classifying them into particular roles." Yes, but my classification is at least in part according to Vietnamese sex worker categories of choice (Walters 2013). Second, it does help this analysis enormously to begin with such classification.

Regarding mistress, Griffin (1999:16) considers it simply as "some-one who is having an affair with a man who is married to someone else." In that sense, it is also reminiscent of Japanese geisha spending two, three or more nights a week in a hotel or in a residence he sets up for her, with their *danna*. He also takes the geisha with him on trips and even sires children by her in many cases (Golden 1997). The geisha is in that sense a mistress too.

Nine

Disordered Global Society

We live in a world of extreme inequality the like of which was without familiarity or expectation prior to the Holocene. Then the onset of Very Big Hierarchy (VBH) societies some 5000 years ago (see Section 6) set us upon a path of global exploitation and wealth partitioning that everybody alive today now has to take for granted as the way things are.

For nearly six million years hominids strode the planet in societies with minimal wealth inequality. Then suddenly, catastrophically, with the onset of VBH societies at the eastern end of the Mediterranean, all human societies began to be engulfed in this expanding global form of socio-economic organisation. Historians, economists, philosophers and students of politics would recognise in the nineteenth and early twentieth centuries the latest manifestation of this organisation as the capitalism that has ruled European and global society for some 500 years.

In this Section I deal briefly with this issue of modern inequality. Much is now being written on the topic, so my contribution is small in aim: to show how a statistical approach can help us describe and account for much of the pattern we see.

Modern global inequality is often described by the kinds of frequency distributions we have encountered in this book. I give one example in Figure 2 below. It shows the frequency distribution of household incomes in Australia during 2013-24 (Australian Bureau of Statistics [ABS] 2015). Notice that as would be expected for a log normal type distribution some 60% or so of incomes lie in the modal clump between $300 through about $1200. Much smaller and decreasing percentages of households have incomes larger than that while a small percentage has incomes less than $300.

From data like this, with strong samples, analysts can then deploy various metrics to explore the distribution further. Median household income is a measure we often hear spoken about in financial news broadcast on radio or television or presented in print media. Sometimes means rather than medians are used as the measure of central tendency. Economists from the Australian Bureau of Statistics – the source of Figure 2 – tend to use a measure known as the Gini coefficient (mentioned already in Section 2) to examine the spread of such a distribution. This measure was used in the paper from which Figure 2 is taken (ABS 2015).

Inequality can be assessed or described by measures such as the Gini, or it can be discussed in a policy setting through more political or sociological concepts such as fairness, living wage, social wage, poverty indices, and so on. Description by means of frequency distribution information allows us to build some ideas according to the foundational laws set out in Section 2 and explored briefly in Sections 4 and 6.

Below I will simply make some brief observations in this regard.

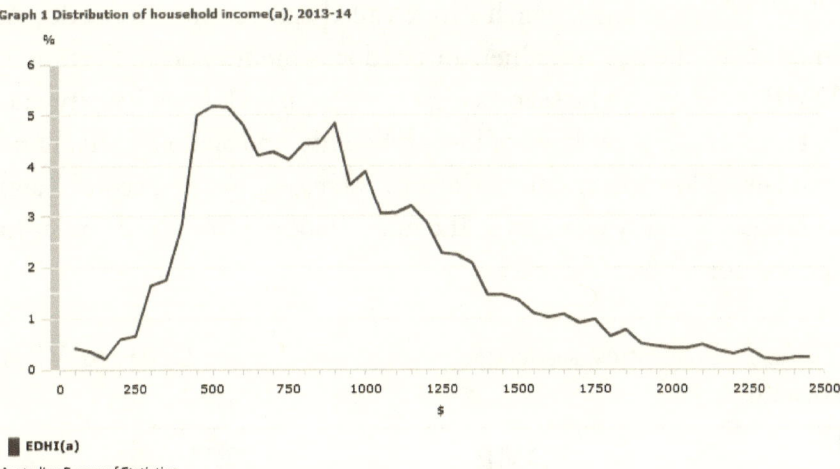

Graph 1 Distribution of household income(a), 2013-14

EDHI(a)

Australian Bureau of Statistics

© Commonwealth of Australia 2017.

Figure 2. Frequency distribution of household incomes in Australia 2013-14.

Domestic Mode of Production (DMP) societies (see Section 6) were small scale. Social and cultural control must have been effective as such social forms remained stable for nearly 6 million years. Control over next and subsequent generations must have meant little to no political dissent on any scale that would have led to breakaway organisations. This claim may have even applied to belief systems.

The Holocene saw the onset of Very Big Hierarchy (VBH) societies which incorporated powerful kings with policing and military support, together with newly established bureaucracies and belief systems. Things taken for granted today such as taxation arrived; social categories such as serfdom and slavery took on higher profile roles. A small set of social functionaries controlled resource distribution, wealth acquisition and social power.

Think of this change, which I have called catastrophic in Section 6, in terms of the gas containers or paddocks mentioned in Section 2. DMP societies can be seen as highly ordered, with low diversity, and in terms of our great laws, of low probability, and again by Shannon's first law, of low information. Entropy increases (by the Second Law) as human sociality changes in the mid Holocene from more order to more disorder (see Table 2).

Table 2. Properties associated with change from DMP to VBH societies.

	DMP	VBH
	entropy increasing →	
order	ordered	disordered
diversity	low	high
probability	low	high
information	low	high

At first take this may perhaps seem to be contradicted by features of modern society that reflect political claims for public order, civil order, and the power structures of the wealthy, their governments, and their implementation arms such as military, police, and bureaucracy. In modern global capitalism barons like Murdoch or the Rothschilds seek order so they can have an established position of supreme power and vast wealth acquisition which go relatively unchallenged. But despite their control over resources and wealth – and their demands, via their governments and their media, for law and order - they are actually agents of disorder.

While this may seem counterintuitive, it needs to be thought about in terms of our fundamental laws, rather than through the clichés of public discourse. Take Murdoch as example. His vast power and wealth, his ownership of resources, media empires, his monolithic control of or at least heavy influence on right wing political parties on several continents, significantly contributes to a public disorder in current Western democracies (English speaking anyhow). While his media and his governments talk order, he is at work undermining that very order. The powerful and wealthy like him want order for *us*, while they remain on an ever freer leash – courtesy of legislation enacted by their GOP, Tory or Liberal Party governments - to get away with whatever they can. They agitate at the van of historical action which is no less than a Radical Right Wing Revolution in global VBH society.

So where do VBH societies go from here? My suspicion is for radical global corporatism leading to massive disenfranchisement, serfdom, or even slavery. The Second Law shows us that while there may be pockets of localised order (entropy decreasing), the global situation more than compensates for this with disorder the lived reality. It looms as the legacy we bequeath our children and their children.

In Walters (2013, 2015) I suggested that VBH society as a phenomenon of cultural evolution is unstoppable. Table 2 reinforces that view. It shows that following 6 million years of stable ordered DMP society, we now not only live through a radically disordered society, but that it is one that follows from the Second Law of thermodynamics. The increase in entropy leads to concomitant changes whereby modern society contains far more information – which when controlled by a few becomes a very powerful weapon – and a higher probability. Which is not to say VBH society was inevitable – far from it – or that it cannot be controlled by political will. Some would hope it can. But it is a social and cultural phenomenon that brings with it grave danger for the future, for society, and for the planet. As we are human and we can be agents of change, perhaps there will political will for the good life for all.

Ten

Bayesian Analysis

Bayes theorem is a well known formula for analysing what are called conditional probabilities. That is, factors that depend on some other condition. Do I take my umbrella today, given that rain is forecast? Statisticians use Bayesian analysis to cope with such conditional dilemmas. It is about making rational choices under uncertainty. In a stunningly counterintuitive fashion, Bayes theorem suggests that the extent to which the probability of the factor we are concerned with (do I take my umbrella) in relation to some event (rain) depends on its opposite or inverse. That is, the probability that there will be rain given that I take my umbrella and that it will rain given that I do not take my umbrella.

In other words it all concerns the relationship between what is called the prior, or prior probability, and the posterior probability. The former, as the name suggests, is the information we have about a situation (jargon: parameters), while the latter gives us a probability for this combined with any new information (measurements) we may have more recently come by.

Bayesian inference or analysis is a very powerful tool for making decisions under uncertainty. It has lent a logical power to decision theory, making it one of the central utilities deployed in statistical approaches used in science, engineering, economics and indeed more recently, anthropology.

In this section I provide two examples of decision making under uncertainty using the analysis that follows from Bayes theorem. One – very brief - is from radiocarbon dating in archaeology and provides short comments on the basics involved, while the other – slightly less brief - hails from an actual data context in ethnology, whence my own sex work research is visited again.

For some decades now Bayesian analysis has been widely used as a method to calibrate radiocarbon dates. The analysis uses information from new measurements together with information from the ^{14}C calibration curve (Bronk Ramsey 2009:337). Widely available software has been developed for handling the computations involved. Bronk Ramsey (2009) provides a key background to the methods involved, an example of the deployment of a program (OxCal), as well as details of the underlying mathematical formalism.

When measurements are made regarding the timing of an event some dates are much more likely than others. This is usually expressed in terms of a probability density function. Bronk Ramsey (2009) gives an example of a calendrical date – reign of William the Conqueror – in terms of a simple uniform probability density function based on the dates of his conquest and death. When radiocarbon dating is required, it is not quite so straightforward. But Bayesian inference comes to the rescue.

Because of the analysis underpinning this dating, namely an isotopic ratio, calibration is required to convert the result into an age. This procedure is essentially Bayesian in form. Many models are available, ranging from very simple to more complex, as are software diag-nostics for exploring their implications (Bronk Ramsey 2009:358). Examples of such models are discussed in for example, Veth *et al.* (2017).

Using Bayesian techniques a "surprising amount" of information can be extracted from such analyses allowing a much improved interpre-tation of the data than was formerly possible by merely eye balling a set of calibrated dates (Bronk Ramsey 2009:358).

Now I turn to the slightly more detailed second example. In Walters (2013) I performed a Bayesian analysis of decision making under uncertainty to suggest that the high probability of prostitution ad-vancing a better lifestyle for families in Vietnam at the turn of the century made it the sensible, as well as moral, thing to do. I asked the simplest question about prior and posterior probabilities for Vietnamese families seeking and gaining a better lifestyle. That is, using Bayes theorem to allow us to see, with a prior probability of a better life for Vietnamese families, what is the posterior probability of a better life if the family sends a daughter to prostitution.

To perform such an analysis required estimates of certain variables. For what became Walters (2013) I collected data beginning in 1998. So I chose to use some social indicators relevant to the state of Vietnamese society at that time. Le Thi wrote a lengthy study on the Vietnamese family, published in 1999, citing survey data from the years of the 1990s and before. One such survey asked respondents if they thought they now had a better life than they did three years previously. The survey involved some 91,000 households across the "whole country," and 51.71% said they now had better lives.

My question concerned the approximately 50% of Vietnamese

family households who responded that they have better lives now than in some time past. And because my study was about prostitution I wanted to explore the probability that Vietnamese have a better life if they send a daughter to prostitution. In other words, I was asking the question, what are the chances that Vietnamese family households will have better lives if they send a daughter to prostitution?

It turns out that the prior probability of a better life given by the estimate 51.71% (Le Thi 1999:67) shifts in a positive direction such that if a family sends a daughter to prostitution, there will be an 83.72% chance they will attain a better life. At the outset, using the estimate from Le Thi's survey data, there was basically a coin toss probability of people judging their lives to be better in the most recent period (roughly 50:50). A smidgen over half saw their lives as being better. With prostitution figured in to the mix, if the family sent a daughter to prostitution, this increases markedly the chance that their lives would be better to greater than 4 in 5 (that is, >80%). Full details of the Bayesian analysis are given in Walters (2013:401-6).

The clear conclusion from this analysis is that if a) you want a better life for your family, and b) you have daughters in prostitution or available for prostitution, then send them there. To not do so is equivalent to keeping your family locked in a 1 in 2 chance of a better life. To send them there, to have them do prostitution, is to raise this to a 4 in 5 chance, a greater than 80% chance, that your family will attain a better life. This is the profound and compelling reason why – during that era at least - prostitutes should do prostitution in Vietnam.

These two examples, briefly presented here, are intended to merely give readers knowledge that such statistical techniques are available and able to be put to good use in our discipline. Decision theory is a powerful tool for both big and small samples. It allows more mature and sophisticated interpretations of data sets concerning the occurrence of conditional events in the face of uncertainty.

Eleven

The Boas-Lewontin Law: A Proposal

For all that he talked – at least in the early part of his career - about laws and developing the laws of history, Franz Boas was sceptical of deduction, prior classification and laws themselves. Over time he "became more and more sceptical as to the possibilities of … derivation of scientific law" (Stocking 1974:15). Such are the contradictions and antinomies in the thoughts and writings of the great man. And of course we have to allow him to change with age and experience.

The more we read him in the original, the more we see a man of the nineteenth century. Not only was he a scholar of his times, but he thought and worked steadfastly within the bounds of the concepts of his times. Race was big on his agenda, perhaps his prime concern, and ever present in his writings. Yet he saw with great insight what so many of his nineteenth century colleagues failed to see: the cultural artifice of the very concept race, and the care with which it needs to be applied, if at all. This last caveat reinforces the complexities associated with Boas, marking him as very much a man of the twentieth century also.

In addition, he was, at least in the earlier part of his career, a man concerned with finding laws for social and cultural sciences. And such is the irony bequeathed to both history and culture, that while acknowledging his later disillusion with the possibility of finding such laws, I am here to suggest a new law of science based upon the findings of his research, and bearing his name.

Amidst his voluminous quantitative work on physical form, morphology, growth, heredity, and so on, Boas also included a question which asked simply: are the differences between so-called races really all that important in human diversity?

He argued time and again, and was able to show empirically that the greatest proportion of difference in human bodily morphology occurred within-groups. Much less difference was accounted for by differences between groups. That is to say, if we are dealing with supposed racial differences, far less difference occurred between the so-called races than occurred within the particular races themselves. The so-called Aryans were more diverse among themselves than they were different to Jews, or Iberians, for example.

Amazingly, the results and profound conclusions of this finding have been largely ignored by anthropology, and by science in general. Perhaps once Hitler was defeated, many scholars could see little interest in such study. (But ask anyone involved in the Civil Rights days of the US South if there was little interest.) Boas had plied his politics against the Nazis, and found one way for his science to reject their ideological clamour as well.

Such was the ignore of this work that when Richard Lewontin (1972, 1974) came to the same conclusion half a century later, with quantitative findings from studies on the fruit fly *Drosophila melanogaster*, he too appears to have been completely unaware that Boas had found exactly the same thing for humans all those decades before. In the 1972 paper he actually referred to a set of other studies that have, since his original findings, shown the result to be widely confirmed in many animal groups, including "man" (Lewontin 1972:382). But still no mention was given to Boas' discovery.

Boas' work was able to withstand not only the empirical tests of time, but the withering political opposition he faced among nineteenth century racialists in both Europe and North America. The Boas breakthrough was sufficiently important in its time, but it is also of continuing significance in a world harmed by racism and bigotry. Hence I think it high time to revive it, explaining its ground breaking importance while linking it to the geneticist who recapitulated it, in some sense, in more modern times. I am therefore going to propose a new law of science named after him: the Boas-Lewontin Law. (Well, half named after him, obviously.)

It will state simply that the physical diversity (or amount of difference) within a given group of humans or any other living beings is never less than the difference between that group and any other given group.

As it was originally worded by the man himself: "the differences between different types of man are, on the whole, small as compared to the range of variation in each type" (Stocking 1968:192).

He stated this many times in slightly different wordings:

- the "indefiniteness of distinctions between different types is due to the variability of the types … and to the comparatively small differences between the types" (Boas 1922:33);

- "the differences between different types of man are, on the whole, small as compared to the range of variation in each type" (Boas 1922:94);

- "the wide range of individual variability in each race" rendered differences between races "insignificant" by comparison (Boas 1974c:314);

- "every racial group consists of a great many family lines which are distinct in bodily form. Some of these family lines are duplicated in neighbouring territories and the more duplication exists the less is it possible to speak of fundamental racial characteristics ... differences between the family lines belonging to each larger area are much greater than the differences between the populations as a whole" (Boas 1940:5);

- and again, from a slightly different context, (and in what might seem to our modern political correctness, slightly clunky expression): "half-breeds differ among themselves more than do the pure races" (Boas 1974e:195).

As noted, there is a second, post-hyphen name there too. Hence my claim for a new law also requires me to talk briefly about Richard Lewontin, the second part of the tag team title I have chosen. Lewontin, is an evolutionary biologist and population geneticist, whose research and writing have spread across a broad range of ecological and environmental biology as well as eco-politics (see for example Levins & Lewontin 1985). Lewontin (1972, 1974) initially used experimental data concerning flies of the genus *Drosophila*, then later, using data from humans, discovered the exact same pattern Boas has discovered fifty years before. His results showed that 85% of difference was found within local geographic groups, while differences attributable to so-called races accounted for only 7.5 – 15%. That is, diversity within groups was greater than or equal to the diversity (or difference) between any given pair of groups. Or to state it slightly more accurately, the differences found between groups accounted for the smallest contribution to overall difference in a sample while within-group difference accounted for by far the largest part of overall difference.

As I said it seems Lewontin had never heard of Boas' work, or perhaps even of the man himself. Hence his discovery was completely independent, basically coming to be regarded, at least by fellow biologists, as the first unveiling of this pattern. Boas' role in the earlier discovery had once again gone unnoticed. Twenty-five or so years ago I learned about it from Stocking (1968), whereupon I soon began teaching it to my third year undergraduate students in the context of it being a significant finding by the great man that almost nobody knew about. Stocking himself was, I think, largely unaware of its more general evolutionary significance, and most certainly unaware – at least according to his lack of reference to it – of Lewontin's later work in genetics and biology.

Physical diversity can be considered in terms of limb or bone measurements, such as the data Boas (1912) compiled in New York City over a hundred years ago, or in terms of genetic characters, such as those studied by Lewontin 50 or 60 years later. I am well aware that many other physical measures may also be used, biochemical features such as shared DNA, anatomical features such as blood groups, external features such as fingerprint or skin patterns, etc.

Of course caveats of sample size, temporal contiguity, and geographical proximity must be taken into account. No one would be surprised, for example, that a group of 3 million year old Australopithecines are going to be far more different from a group of modern Africans than either is diverse within its group. So things must be kept sensible when setting the boundary conditions of the law. But this is no different from demands that Newton's Laws of Motion will not apply to objects approaching closely the velocity of electromagnetic radiation in a vacuum (speed of light). That's the boundary condition that applies to Newton's great laws. Beyond it relativistic considerations apply.

Such a law as I propose is and has been empirically induced – to use the jargon briefly – rather than deductively derived. This means it holds status after the manner of Newton's Laws, inasmuch as those laws derived from the evidence of Galileo and others; Maxwell's Laws, which derived in major part from the empirical findings of Faraday; many other laws in basic chemistry and physics, such as Boyle's Law, Charles' Law, Stokes' Law, etc. have this same quality.

There have been philosophical style counter claims and refutations of Lewontin's work, accusing him a being too political, allowing his personal politics to influence the assumptions of his science. (Is that a fairly standard mainstream response to findings that upset conservative apple carts?) But I know of no acceptable or creditable experimental study that has ever violated the Boas-Lewontin results. Hence my claim for it to qualify as a new and genuine law of science.

The importance of Boas' discovery is self-evident to anyone familiar with the later biological findings. It is for this reason the independent discovery by these two scientists, separated by both time and disciplinary insularity, that I choose to name the law they discovered after both of them, chronological precedent determining name order. This great law was a triumph in the war on racism, or at least should have been were evidence ever to count over ideological politics. It was also a triumph, in more prosaic academic ways, for statistical anthropology.

Franz Boas, for all his twentieth century influence, academic power and global scholarly reputation, remained in some key intellectual dimensions, very much a creature of the nineteenth century. It needs to be remembered that the turn of the century exactly bisected his life: he was forty-two years old then, and died in 1942 at age 84. Hence he was no doubt well set in many of his intellectual ways by the time the twentieth century opened. The most obvious and contradictory of these was race.

His work was known and admired by the likes of Tylor and Levi Strauss, well known, if less admired, by the likes of Radcliffe-Brown, and well known and loathed by the Nazis who burned his books and forced the University of Kiel to revoke his doctorate. He was, as said earlier, the most important figure in the world, either academically or politically, in the argument for evidence-based rejection of the idea that one race might be superior to another race or races.

Yet in contradictory fashion, he remained harnessed to the idea of race. In all his writings we see him taking race for granted, every bit as much as any other scholar of politician concerned with the issue. What set him apart, as I have said, was his use of physical evidence, empirical evidence, cultural evidence to show that no one race held necessary superiority over others. Should we care to provide a modern interpretation of his findings, we could justifiably claim that his arguments went beyond this, all the way to a very demonstration that race was therefore an unnecessary and burdensome cultural category that the world would do well to abandon, both academically and politically.

Twelve

Conclusions

Franz Boas showed us the way. He was on the money introducing a statistical anthropology while Einstein and Gibbs were creating statistical physics. It is not his fault that anthropology abandoned it while students of physics, mathematics, engineering and biology launched their work into significant new fields such as quantum mechanics, information theory, and mathematical ecology. I argue it may be time for anthropology to revisit this endeavour. Our Boas ranks as one of the great empiricists of late nineteenth and early twentieth century science, a scholar whose reputation deserves to sit comfortably alongside luminaries of empirical discovery such as his contemporaries Faraday and Rutherford. We would do well to follow in his footsteps.

I thus argue that our profession can usefully return to a statistical anthropology. Not every anthropologist, mind, nor in all of her research endeavours. But each to their own. Where the study is suited, then do it and do it well. For those inclined to numbers look beyond your SPSS buttons, think beyond means, standard deviations and *t* tests. Teach statistical anthropology and teach it as I have advocated here, with foci on distributions, on what makes up, for example, a normal curve, or a rank-size distribution, on diversity, on probability, and on decision making. Don't be afraid of laws and law-like generalisations.

Looked at through entropy, disorder and diversity, statistical anthropology allows us to ponder the future of disordered global society. It locates us therefore alongside politics, sociology, history and economics as potentially making a contribution to theories about the future of Very Big Hierarchy societies (of which capitalism is the most recent venture), of our species, and of our planet.

However, for all its theoretical and intellectual benefits, a statistical anthropology also still anchors us firmly and soundly in very practical outcomes of relevance to both society in general and to the peoples we study. One example demonstrated here concerns sex workers and their sex industry. Theoretical outcomes of some significance have startling pragmatic consequences for practitioners, their supporters, legislators, police forces and the law – and, for would-be abolitionists. There are important practical implications to be understood by ideological opponents of sex work and sex industries.

Then, also as an example of a practical outcome, are the concerns with how we deal with race.

The Boas-Lewontin Law, if I have it right, suggests we would do well as scientists and as politicians, or at least as advocates, to abandon the concept of race. It holds little by way of useful explanatory value, and almost nothing in terms of understanding significant factors of human diversity and difference. Politics would be a far nicer preoccupation without it. So would anatomy. (It and old-time physical anthropology still have things to answer for.) Recall the words of James Clifford from Section 1: "difference is no longer a stable, exotic otherness; self-other relations are matters of power and rhetoric rather than of essence" (Clifford 1988:14).

We could do worse than make sure that forever we turn our collective backs on rigid racial categorisation, on bigotry, on any form of Right Wing extremism, on the Nazi path, on the Stalinist path, and follow instead the roads trod by decent tolerant inclusive global citizens and democrats. Go with diversity, not racial categories. Make decisions acknowledging uncertainty; avoid the trap of thinking ideology equates with or leads to certainty. Make use of the vast amount of data and information available in modern disordered society; don't deny it.

No more book burning, real or metaphorical. No more denial of science. What's needed is a return to pre-post-truth. Or perhaps we push on through to post-post-truth; but that brings us anyway to a place equivalent to pre-post-truth.

References

Allen, H. 1972. Where the Crow Flies Backwards: Man and Land in the Darling Basin. Unpublished PhD thesis, Australian National University.

Allen, John S. 1989. Franz Boas's physical anthropology: the critique of racial formalism revisited. *Current Anthropology* 30(1):79-84.

Alexsander, Igor 2003. Understanding information, bit by bit: Shannon's equations. In Graham Farmelo (Editor) *It Must Be Beautiful: Great Equations of Modern Science*, pp.213-230. London: Granta.

Andaya, Barbara Watson 1998. From temporary wife to prostitute: sexuality and economic change in early modern Southeast Asia. *Journal of Women's History* 9(4):11-35.

Anon. 1988. *Nam: The Vietnam Experience 1965-1975*. London: Hamlyn.

Anon. 1998. China's new sex trade booming. *Northern Territory News*, Monday 3 August, p.10.

Arnold, Vladimir I. 2004. *Catastrophe Theory*. Third Edition. Berlin: Springer-Verlag.

Atkins, Peter 2010. *The Laws of Thermodynamics: A Very Short Introduction*. Oxford: Oxford University Press.

Australian Bureau of Statistics 2015. Summary indicators of income and wealth distribution. 6553.0 – Survey of Income and Housing, User Guide, Australia, 2013-14. Canberra: ABS.

Balme, J. 1983. Prehistoric fishing in the lower Darling, western New South Wales. In C.Grigson & J.Clutton-Brock (Editors) *Animals and Archaeology: 2. Shell Middens, Fishes and Birds*. BAR International Series 183:19-32.

Blackwell, A. 1980. Oh, I Do Like to be Beside the Seaside: Results from the Bowen Island excavation. Unpublished BA(Hons) thesis, Australian National University.

Boas, Franz 1912. *Changes in Bodily Form of Descendants of Immigrants*. New York: Columbia University Press. [Elibron Classics Replica Edition].

Boas, Franz 1922. *The Mind of Primitive Man*. New York: Macmillan [Forgotten Books Classic Reprint Series Edition].

Boas, Franz 1940. *Race, Language and Culture*. New York: Macmillan [Forgotten Books Classic Reprint Series Edition].

Boas, Franz. 1973 [original 1896]. The limitations of the comparative method of anthropology. In Paul Bohannan & Mark

Glazer (Editors) *High Points in Anthropology*, pp. 84-92. New York: Knopf.

Boas, Franz 1974a [1904]. The history of anthropology. In Stocking, George W. Jr (Editor) *A Franz Boas Reader: The Shaping of American Anthropology, 1883-1911*, pp.23-36. Midway Reprint. Chicago: University of Chicago Press.

Boas, Franz 1974b [1898]. The Jesup North Pacific Expedition. In Stocking, George W. Jr (Editor) *A Franz Boas Reader: The Shaping of American Anthropology, 1883-1911*, pp.107-16. Midway Reprint. Chicago: University of Chicago Press.

Boas, Franz 1974c [1906]. The outlook for the American Negro. In Stocking, George W. Jr (Editor) *A Franz Boas Reader: The Shaping of American Anthropology, 1883-1911*, pp.310-16. Midway Reprint. Chicago: University of Chicago Press.

Boas, Franz 1974d [1902]. Anthropological instruction in Columbia University. In Stocking, George W. Jr (Editor) *A Franz Boas Reader: The Shaping of American Anthropology, 1883-1911*, pp.290-3. Midway Reprint. Chicago: University of Chicago Press.

Boas, Franz 1974e [1894]. The anthropology of the North American Indian. In Stocking, George W. Jr (Editor) *A Franz Boas Reader: The Shaping of American Anthropology, 1883-1911*, pp.191-201. Midway Reprint. Chicago: University of Chicago Press.

Bowdler, S. & H.Lourandos 1982. Both sides of Bass Strait. In S.Bowdler (Editor) *Coastal Archaeology in Eastern Australia*, pp.121-32. Department of Prehistory, Australian National University.

Bronk Ramsey, Christopher 2009. Bayesian analysis of radiocarbon dates. *Radiocarbon* 51(1):337-60.

Casteel, R.W. 1972. Some archaeological uses of fish remains. *American Antiquity* 37:404-19.

Casteel, R.W. 1974. A method for estimation of live weight of fish from the size of skeletal elements. *American Antiquity* 39:94-8.

Casteel, R.W. 1976. *Fish Remains in Archaeology and Paleo-Environmental Studies*. London: Academic Press.

Chapkis, Wendy 1997. *Live Sex Acts: Women Performing Erotic Labour*. London: Cassell.

Clifford, James 1988. *The Predicament of Culture: Twentieth-Century Ethnography, Literature, and Art*. Cambridge, Mass.: Harvard University Press.

Coleman, J. 1978. The Analysis of Vertebrate Faunal Remains from four shell middens in the Lower Macleay River District. Unpublished BA(Hons) Thesis, University of New England, Armidale.

Coleman, J. 1980. Fish bones for fun and profit. In I.Johnson (Editor) *Holier Than Thou: Proceedings, Conference on Australian Prehistory*, pp.61-75, Department of Prehistory, Australian National University.

Colley, Sarah M. 1983. The Role of Fish Bone Studies in Economic Archaeology: with special reference to the Orkney Isles. PhD Thesis, University of Southhampton.

Dutton, Michael 1998. *Streetlife China*. Cambridge: Cambridge University Press.

Einstein, Albert 1998a [1905a]. A new determination of molecular dimensions. In John Stachel (Editor) *Einstein's Miraculous Year: Five Papers That Changed the Face of Physics*, pp.45-69. Princeton: PrincetonUniversity Press.

Einstein, Albert 1998b [1905b]. On the motion of small particles suspended in liquids at rest required by the molecular-kinetic theory of heat. In John Stachel (Editor) *Einstein's Miraculous Year: Five Papers That Changed the Face of Physics*, pp.85-98. Princeton: PrincetonUniversity Press.

Evans, Harriet 1997. *Women and Sexuality in China: Dominant Discourses of Female Sexuality and Gender since 1949*. Cambridge: Polity.

Farmelo, Graham 2009. *The Strangest Man: The Hidden Life of Paul Dirac, Quantum Genius*. London: Faber & Faber.

Fermi, Enrico. 1956 [1936]. *Thermodynamics*. New York: Dover.

Galison, Peter 2003. The sextant equation: $E = mc^2$. In Graham Farmelo (Editor) *It Must Be Beautiful: Great Equations of Modern Science*, pp.68-86. London: Granta.

Geertz, Clifford 1973. *The Interpretation of Cultures*. New York: Basic Books.

Golden, Arthur 1997. *Memoirs of a Geisha*. London: Vintage.

Greene, Graham 1973 [1955]. *The Quiet American*. Harmondsworth: Penguin.

Griffin, Victoria 1999. *The Mistress: Histories, Myths and Interpretations of the 'Other Woman.'* London: Bloomsbury.

Hamley, J.M. 1975. Review of gillnet selectivity. *Journal of the Fisheries Research Board of Canada* 32:1943-1969.

Harris, Marvin 1968. *The Rise of Anthropological Theory*. New York: Crowell.

Harris, Marvin 1979. *Cultural Materialism: The Struggle for a Science of Culture*. New York: Vintage.

Hershatter, Gail 1997. *Dangerous Pleasures: Prostitution and Modernity in Twentieth-Century Shanghai*. Berkeley: University of California Press.

Holling, C.S. 2010a. Resilience and stability of ecological systems. In Lance H.Gunderson, Craig R.Allen & C.S.Holling (Editors) *Foundations of Ecological Resilience*, pp. 19-49. Washington: Island Press.

Holling, C.S. 2010b. Engineering resilience versus ecological resilience. In Lance H.Gunderson, Craig R.Allen & C.S.Holling (Editors) *Foundations of Ecological Resilience*, pp. 51-66. Washington: Island Press.

Jamieson, Neil L. 1993. *Understanding Vietnam*. Berkeley: University of California Press.

Johnson, Gregory A. 1987. The changing organization in Uruk administration on the Susiana Plain. In F. Hole (Editor) *The Archaeology of Western Iran*. Washington DC: Smithsonian Institution Press.

Kefous, K.C.1977. We Have a Fish with Ears, and Wonder if it is Valuable? Unpublished BA(Hons) Thesis, Australian National University.

Kempadoo, Kamala 1998. Introduction: globalizing sex workers' rights. In Kamala Kempadoo & Jo Doezema (Editors) *Global Sex Workers: Rights, Resistance, and Redefinition*, pp.1-28. New York: Routledge.

Kuper, Adam 1983. *Anthropology and Anthropologists: The Modern British School*. London: Routledge & Kegan Paul.

Lapin, L.L. 1973. *Statistics for Modern Business Decisions*. New York: Harcourt Brace Jovanovich.

Le Thi 1999. *The Role of the Family in the Formation of Vietnamese Personality*. Hanoi: The Gioi Publishers.

Le Thi Quy 1993. Some ideas about prostitution in Vietnam. Paper presented at the Conference on Joining Forces to Further Shared Visions, Washington DC, USA, October 20-24, 1993.

Leach, B.F. 1979. Fish and crayfish from the Washpool midden site, New Zealand: their use in determining season of occupation and prehistoric fishing methods. *Journal of Archaeological Science* 6:109-26.

Levins, Richard & Richard Lewontin 1985. *The Dialectical Biologist*. Cambridge, Mass.: Harvard University Press.

Lewontin, R.C. 1972. The apportionment of human diversity. *Evolutionary Biology* 6:381-98.

Lewontin, R.C. 1974. *The Genetic Basis of Evolutionary Change*. New York: Columbia University Press.

MacArthur, John W. 1975. Environmental fluctuations and species diversity. In Martin L. Cody & Jared M.Diamond (Editors) *Ecology and Evolution of Communities*, pp.74-80. Cambridge, Mass.: Belknap of Harvard.

Magurran, Anne E. 2004. *Measuring Biological Diversity*. Malden, MA: Blackwell.

May, Robert M. 1975. Patterns of species abundance and diversity. In Martin L. Cody & Jared M.Diamond (Editors) *Ecology and Evolution of Communities*, pp.81-120. Cambridge, Mass.: Belknap of Harvard.

Micollier, Evelyne 2004. Social significance of commercial sex work: implicitly shaping a sexual culture? In Evelyne Micollier (Editor) *Sexual Cultures in East Asia: The Social Construction of Sexuality and Sexual Risk in a Time of AIDS*, pp.3-22. London & New York: Routledge Curzon – IIAS Asian Studies Series.

Moore, Walter 1989. *Schrödinger: Life and Thought*. Cambridge: Cambridge University Press.

Nahin, Paul J. 2002 [1987]. *Oliver Heaviside: The Life, Work, and Times of an Electrical Genius of the Victorian Age*. Baltimore: Johns Hopkins University Press.

Nguyen-Vo Thu-Huong 2008. *The Ironies of Freedom: Sex, Culture, and Neoliberal Governance in Vietnam*. Seattle: University of Washington Press.

O'Keefe, Brendan G. with F.B.Smith 1994. *Medicine at War: Medical Aspects of Australia's Involvement in Southeast Asia 1950-1972*. St Leonards, NSW: Allen & Unwin, in association with the Australian War Memorial.

Pais, Abraham 2008. *Subtle is the Lord: The Science and the Life of Albert Einstein*. Oxford: Oxford University Press.

Pateman, Carole 1988. *The Sexual Contract*. London: Polity Press.

Penrose, Roger 2008. Foreword, in Abraham Pais *Subtle is the Lord: The Science and the Life of Albert Einstein*, pp.vii-x. Oxford: Oxford University Press.

Pope, J.A. 1966. *Manual of Methods for Fish Stock Assessment. Part 3. Selectivity of Fishing Gear*. FAO Fisheries Technical Paper No.41. Rome: Food and Agriculture Organization of the United Nations.

Pope, J.A., A.R.Margetts, J.M.Hamley & E.F.Akuz 1975. *Manual of Methods for Fish Stock Assessment. Part 3. Selectivity of Fishing Gear*. FAO Fisheries Technical Paper No.41, Revision 1. Rome: Food and Agriculture Organization of the United Nations.

Poston, Tim & Ian Stewart 1996 [originally 1978]. *Catastrophe Theory and Its Applications*. New York: Dover.

Radcliffe-Brown, A.R. 1931. *The Social Organisation of Australian Tribes*. Oceania Monographs. Melbourne: Macmillan.

Radcliffe-Brown, A.R. 1979 [1952]. *Structure and Function in Primitive Society*. London: Routledge & Kegan Paul.

Reid, Anthony 1993a. *Southeast Asia in the Age of Commerce 1450-1680: Volume One: The Lands Below the Winds*. New Haven: Yale University Press.

Reid, Anthony 1993b. *Southeast Asia in the Age of Commerce 1450-1680: Volume Two: Expansion and Crisis*. New Haven: Yale University Press.

Royce, W.F. 1972. *Introduction to the Fishery Sciences*. New York: Academic.

Sahlins, Marshall 1974. *Stone Age Economics*. London: Tavistock.

Schrödinger, Erwin 1989 [1952]. *Statistical Thermodynamics*. New York: Dover.

Scott, George Ryléy 1996 [1968]. *The History of Prostitution*. Twickenham, Middlesex: Senate.

Service, Elman R. 1971. *Cultural Evolutionism: Theory in Practice*. New York: Holt, Rinehart & Winston.

Shannon, Claude E. & Warren Weaver 1998 [1949]. *The*

Mathematical Theory of Communication. Urbana: University of Illinois Press.

Sheehan, Neil 1990. *A Bright Shining Lie: John Paul Vann and America in Vietnam.* London: Picador.

Stocking, George W. Jr 1968. The critique of racial formalism. In *Race, Culture, and Evolution: Essays in the History of Anthropology*, pp.161-94. New York: The Free Press.

Stocking, George W. Jr 1974a. Introduction: The basic assumptions of Boasian anthropology. In Stocking, George W. Jr (Editor) *A Franz Boas Reader: The Shaping of American Anthropology, 1883-1911*, pp.1-20. Midway Reprint. Chicago: University of Chicago Press.

Stocking, George W. Jr 1974b. Anthropology and society. In Stocking, George W. Jr (Editor) *A Franz Boas Reader: The Shaping of American Anthropology, 1883-1911*, pp.307-09. Midway Reprint. Chicago: University of Chicago Press.

Stocking, George W. Jr 1974c. The propagation of anthropology. In Stocking, George W. Jr (Editor) *A Franz Boas Reader: The Shaping of American Anthropology, 1883-1911*, pp.283-6. Midway Reprint. Chicago: University of Chicago Press.

Stocking, George W. Jr 1974d. Basic anthropological viewpoints. In Stocking, George W. Jr (Editor) *A Franz Boas Reader: The Shaping of American Anthropology, 1883-1911*, pp.57-9. Midway Reprint. Chicago: University of Chicago Press.

Stoler, Ann 1997. Educating desire in colonial Southeast Asia:

Foucault, Freud, and imperial sexualities. In Lenore Manderson & Margaret Jolly (Editors) *Sites of Desire, Economies of Pleasure: Sexualities in Asia and the Pacific*, pp.27-47. Chicago & London: University of Chicago Press.

Thom, Rene 1989 [originally 1972]. *Structural Stability and Morphogenesis: An Outline of a General Theory of Models.* Westview Press Advanced Book Classics.

Tolman, Richard C. 1979 [1938]. *The Principles of Statistical Mechanics.* New York: Dover.

Tylor, Edward Burnett 1958 [1871]. *Primitive Culture.* New York: Harper & Row.

Veth, Peter, Ingrid Ward, Tiina Manne, Sean Ulm, Kane Ditchfield, Joe Dortch, Fiona Petchey, Alan Hogg, Daniele Questauix, Martina Demuro, Lee Arnold, Nigel Spooner, Vladimir Levchenko, Jane Skippington, Chae Byrne, Mark Basgall, David Zeanah, David Belton, Petra Helmholz, Szilvia Bajkan, Richard Bailey, Christa Plascek & Peter Kendrick 2017. Early human occupation of a maritime desert, Barrow Island, North-West Australia. *Quaternary Science Reviews* 168:19-29.

Walters, Ian 1980. Einstein's philosophy and a social research problem. University of Queensland Anthropology Museum *Occasional Papers in Anthropology* 10:160-166.

Walters, Ian 1987. Another Kettle of Fish: The Prehistoric Moreton Bay Fishery. PhD Thesis, University of Queensland.

Walters, I. 1996. *Meganthropus* and the hominid taxa of Java. *Bulletin of the Indo-Pacific Prehistory Association* 15:229-234.

Walters, I. 2002. Early hominids in Southeast Asia: older, younger, smarter and more. In P.Kershaw, B.David, N.Tapper, D.Penny & J.Brown (Editors) *Bridging Wallace's Line: The Environmental and Cultural History and Dynamics of the SEAsian-Australian-Region*, pp.255-262. Reiskirchen: Catena Verlag GMBH.

Walters, Ian 2013. *Sex Work in Vietnam*. amazon Kindle Direct Publishing [ebook].

Walters, Ian 2014. *The Theory of Relativity*. Denver, Co.: Outskirts Press.

Walters, Ian 2015. *Prostitution: Recent and Unstoppable*. Singapore: Partridge.

Wenke, Robert J., 1999. *Patterns in Prehistory: Humankind's First Three Million Years*. 4th Edition. New York: Oxford.

Wheatcroft, Deborah 2001. Whores Work: Erotic Labour. Stepping Out from Behind the Red Light. BA (Hons) Thesis, Northern Territory University, Darwin.

Williams, Louise 1998. *Wives, Mistresses and Matriarchs: Asian Women Today*. St Leonard's: Allen & Unwin.

Wright, Henry T. 1987. The Susiana Hinterlands during the era of primary state formation. In F. Hole (Editor) *The Archaeology of Western Iran*. Washington DC: Smithsonian Institution Press.

Index

Dinka 8
disorder 22, 23, 32, 91-5, 110, 111
diversity 4, 5, 23, 25, 32, 35, 38, 44, 51, 62, 68, 69, 74, 80, 81, 85, 86, 94, 103, 104, 106, 109-11
Domestic Mode of Production (DMP) societies 53-7, 93-5

Early Modern period 72, 78
Einstein, Albert 1, 2, 11, 12, 21, 109
Elkin, A.P. 67
entropy 2, 20-3, 94, 95, 110
Evans Pritchard, sir E.E. 8

Faraday, Michael 17, 107, 109
faunal analysis (faunal studies) 42, 52
First Law of thermodynamics 20
fishing, fishing implements, fishing gear 44, 50, 51
frequency distributions 10, 19, 25-8, 31, 32, 39-41, 44, 47, 49-51, 58, 60, 92, 93
functionalism 15, 42

Galileo Galilei 17, 107
Gauss, Carl F. 19, 25, 26, 30, 40, 49
Gaussian curve (see Gauss,

normal curve)
Gibbs, Josiah Willard 1, 2, 12, 21, 109
Gini coefficient 28, 92
government 75-7, 79, 84, 86, 87, 94, 95

Harris, Marvin 16, 17, 42
Hershatter, Gail 69, 76-8, 83, 84, 86, 87
Holocene 10, 39, 42, 53, 91, 93, 94

Information 24-6, 37, 45, 61, 68, 78, 85, 92, 94, 95, 97-9, 111
information technology (IT) 24, 68
information theory 19, 23, 27, 109
Inuit 13, 14

Kwakiutl 15

Levi-Strauss, Claude i, ii, 14, 15, 34
Lewontin, Richard 102-7, 110
lognormal distribution 30, 51
logseries distribution 30, 51, 58
Lord Kelvin 19, 20
Lorentz, Hendrik 2, 11, 12
Lorentz-Einstein theory 12

www.ingramcontent.com/pod-product-compliance
Lightning Source LLC
Chambersburg PA
CBHW021957170526
45157CB00003B/1031